Advances in Spatial Science

Springer-Verlag Berlin Heidelberg GmbH

Titles in the Series

C. S. Bertuglia, M. M. Fischer and *G. Preto* (Eds.)
Technological Change, Economic Development and Space
XVI, 354 pages. 1995. ISBN 3-540-59288-1
(out of print)

H. Coccossis and *P. Nijkamp* (Eds.)
Overcoming Isolation
VII, 272 pages. 1995. ISBN 3-540-59423-X

L. Anselin and *R. J. G. M. Florax* (Eds.)
New Directions in Spatial Econometrics
XVX, 420 pages. 1995. ISBN 3-540-60020-5

H. Eskelinen and *F. Snickars* (Eds.)
Competitive European Peripheries
VIII, 271 pages. 1995. ISBN 3-540-60211-9

J. C. J. M. van den Bergh, P. Nijkamp and *P. Rietveld* (Eds.)
Recent Advances in Spatial Equilibrium Modelling
VIII, 392 pages. 1996. ISBN 3-540-60708-0
(out of print)

P. Nijkamp, G. Pepping and *D. Banister* (Eds.)
Telematics and Transport Behaviour
XII, 227 pages. 1996. ISBN 3-540-60919-9

D. F. Batten and *C. Karlsson* (Eds.)
Infrastructure and the Complexity of Economic Development
VIII, 298 pages. 1996. ISBN 3-540-61333-1

T. Puu
Mathematical Location and Land Use Theory
IX, 294 pages. 1997. ISBN 3-540-61819-8

Y. Leung
Intelligent Spatial Decision Support Systems
XV, 470 pages. 1997. ISBN 3-540-62518-6

C. S. Bertuglia, S. Lombardo and *P. Nijkamp* (Eds.)
Innovative Behaviour in Space and Time
X, 437 pages. 1997. ISBN 3-540-62542-9

A. Nagurney and *S. Siokos*
Financial Networks
XVI, 492 pages. 1997. ISBN 3-540-63116-X

M. M. Fischer and *A. Getis* (Eds.)
Recent Developments in Spatial Analysis
X, 434 pages. 1997. ISBN 3-540-63180-1

R. H. M. Emmerink
Information and Pricing in Road Transportation
XVI, 294 pages. 1998. ISBN 3-540-64088-6

P. Rietveld and *F. Bruinsma*
Is Transport Infrastructure Effective?
XVI, 384 pages. 1998. ISBN 3-540-64542-X

P. McCann
The Economics of Industrial Location
XII, 228 pages. 1998. ISBN 3-540-64586-1

L. Lundqvist, L.-G. Mattsson and *T. J. Kim* (Eds.)
Network Infrastructure and the Urban Environment
IX, 414 pages. 1998. ISBN 3-540-64585-3

R. Capello, P. Nijkamp and *G. Pepping* (Eds.)
Sustainable Cities and Energy Policies
XI, 282 pages. 1999. ISBN 3-540-64805-4

M. M. Fischer and *P. Nijkamp* (Eds.)
Spatial Dynamics of European Integration
XII, 367 pages. 1999. ISBN 3-540-65817-3

M. M. Fischer, L. Suarez-Villa and *M. Steiner* (Eds.)
Innovation, Networks and Localities
XI, 336 pages. 1999. ISBN 3-540-65853-X

J. Stillwell, S. Geertman and *S. Openshaw* (Eds.)
Geographical Information and Planning
X, 454 pages. 1999. ISBN 3-540-65902-1

G. J. D. Hewings, M. Sonis, M. Madden and *Y. Kimura* (Eds.)
Understanding and Interpreting Economic Structure
X, 365 pages. 1999. ISBN 3-540-66045-3

A. Reggiani (Ed.)
Spatial Economic Science
XII, 457 pages. 2000. ISBN 3-540-67493-4

D. G. Janelle and *GD. C. Hodge* (Eds.)
Information, Place, and Cyberspace
XII, 381 pages. 2000. ISBN 3-540-67492-6

P. W. J. Batey and *P. Freidrich* (Eds.)
Regional Competition
VIII, 290 pages. 2000. ISBN 3-540-67548-5

B. Johansson, Ch. Karlsson and *R. R. Stough* (Eds.)
Theories of Endogenous Regional Growth
IX, 428 pages. 2001. ISBN 3-540-&7988-X

G. Clarke and *M. Madden* (Eds.)
Regional Science in Business
VIII, 363 pages. 2001. ISBN 3-540-41780-X

M. M. Fischer and *Y. Leung* (Eds.)
GeoComputational Modelling
XII, 279 pages. 2001. ISBN 3-540-41968-3

M. M. Fischer and *J. Fröhlich* (Eds.)
Knowledge, Complexity and Innovation Systems
XII, 477 pages. 2001. ISBN 3-540-41969-1

M. M. Fischer, J. Revilla Diez and *F. Snickars*
Metropolitan Innovation Systems
VIII, 270 pages. 2001. ISBN 3-540-41967-5

L. Lundqvist and *L.-G. Mattson* (Eds.)
National Transport Models
VIII, 202 pages. 2002. ISBN 3-540-42486-1

J. R. Cuadrado-Roura and *M. Parellada*
Regional Convergence in the European Union
VIII, 368 pages. 2002. ISBN 3-540-43242-6

G. J. D. Hewings, M. Sonis and *D. Boyce* (Eds.)
Trade, Networks and Hierarchies
XI, 467 pages. 2002. ISBN 3-540-43087-3

Gündüz Atalik · Manfred M. Fischer
Editors

Regional Development Reconsidered

With 35 Figures
and 32 Tables

Springer

Professor Dr. em. Gündüz Atalik
Istanbul Technical University
Department of Urban and Regional Planning
80191 Taksim, Istanbul
Turkey

Professor Dr. Manfred M. Fischer
Vienna University of Economics and Business Administration
Department of Economic Geography & Geoinformatics
Rossauer Lände 23/1
1090 Vienna
Austria

ISBN 978-3-642-62837-5 ISBN 978-3-642-56194-8 (eBook)
DOI 10.1007/978-3-642-56194-8

Library of Congress Cataloging-in-Publication Data applied for
Die Deutsche Bibliothek - CIP-Einheitsaufnahme
Regional Development Reconsidered / Gündüz Atalik; Manfred M. Fischer ed.. -
Berlin; Heidelberg; New York; Barcelona; Hong Kong; London; Milan; Paris; Tokyo: Springer, 2002
(Advances in Spatial Science)

http://www.springer.de

Originally published by Springer-Verlag Berlin Heidelberg New York in 2002
Softcover reprint of the hardcover 1st edition 2002

Cover design: Erich Kirchner, Heidelberg

SPIN 10878243 42/2202-5 4 3 2 1 0 – Printed on acid-free paper

Preface

In the last few years research on regional development has increased dramatically. Real-world concerns have – to a certain extent – driven this scientific concern of interest. The field has been given a big boost in particular by the process of European integration and the attempt to understand how this deeper integration will work at the regional level.

This volume makes a modest attempt to reconsider the issue of regional development mainly from an European perspective and in the light of the transition of society towards a knowledge-driven economy. It originated from the Thirteenth European Advanced Studies Institute in Regional Science, held in Istanbul, July 2-8, 2000.

In producing the book, as friends and colleagues, we have benefited from the possibility of exchange of ideas and experience. We have also received useful assistance from the referees who have offered observations and advice in their written reports. The soundness of their comments has contributed immensely to the quality of the volume. We should, in addition, like to acknowledge the timely manner in which contributing authors have responded to our requests, and their willingness to follow the stringent editorial guidelines.

We are extremely grateful for the generous support provided by the Department of Urban and Regional Planning at the Istanbul Technical University and the Department of Economic Geography & Geoinformatics at the Vienna University of Economics and Business Administration. We should also like to thank Ingrid Divis, Thomas Seyffertitz and Seda Kundak for their capable assistance in coordinating the various stages of preparation of the book. Finally, we wish to acknowledge the expert editorial assistance provided by Anne Topley. Her care and attention to the linguistic editing of various parts of the book have considerably enhanced the quality of the work presented here.

Istanbul and Vienna, March 2002

Gündüz Atalik
Istanbul Technical University

Manfred M. Fischer
Vienna University of Economics and Business Administration

Contents

PART C: Mobility in Transport Systems in Europe

PART D: Regional Development and Policy Issues

1 Regional Development Reconsidered: A Prologue

Manfred M. Fischer and Gündüz Atalik***
* Vienna University of Economics and Business Administration
Department of Economic Geography & Geoinformatics
** Istanbul Technical University
Department of Urban and Regional Planning

1.1 The Context

It has become increasingly recognised that the location of economic activities cannot be properly understood in isolation from its wider socio-economic context, and, thus, we cannot understand regional development without linking it to the epochal transition to a *knowledge-based* or *learning economy* that we are currently witnessing (OECD, 1996).

This new stage of market-based capitalism is more strongly and more directly rooted in the production, distribution and use of knowledge than ever before (Drucker, 1993). Knowledge creation and diffusion are the key driving forces in the economy, and knowledge is an economic resource in its own right, taking the privileged role that once was accorded to natural resources, even though knowledge *in itself* does not contribute to economic growth. It has to be incorporated into the production of goods and services.

Advances in technological and organisational knowledge have to be absorbed by firms and applied to the production process and organisation of work. This is true irrespective of whether the new knowledge is created externally [such as in universities or research institutions] or internally [such as in the R&D lab of a firm], and irrespective of whether the knowledge advances embody wholly new knowledge or new combinations of existing knowledge. Clearly, the absorptive capacity of organisations varies substantially and this affects their ability to produce innovations. But it is the innovative capability or capacity of organisations that largely determines their competitiveness within the learning economy (Suarez-Villa, 2000). Especially in regions with high factor costs, long term sustainable competitiveness is increasingly related to the ability of firms to improve their performance by means of continuous innovation [in products, processes and work organisation], knowledge creation and learning.

Both, individual and organisational learning are necessary, but of course not sufficient prerequisites for innovations. *Individual learning* involves the acquisition of information and knowledge, understanding and skills by individuals, through formal education and training or learning-by-doing. The result is knowledge accumulated in the heads of people, that is the stock of human capital (Edquist and Rees, 2000). *Organisational learning* depends on individual learning. It can be understood as a process that organisationally amplifies the knowledge created by individuals and crystallises it at the group level through dialogue, discussion, experience sharing, or observation. This process takes place within intra- and interorganisational networks (Nonaka and Takeuchi, 1995). The result of organisational learning is that kind of knowledge capital that cannot easily leave the firm (Edquist and Rees, 2000).

Specific forms of learning and knowledge creation, especially the tacit forms, are both localised and territorially specific. Firms that master non-codifiable knowledge are tied into various kinds of networks, through localised input-output relations, knowledge spillovers and their untraded interdependencies (Storper, 1997).

Intangibles are becoming more and more important in the process of transition, replacing material resources as the predominant source of wealth (Suarez-Villa, 2000). While the intangible assets of firms have become increasingly mobile across national borders (Archibugi, 2001), the location of their production [especially of their tacit elements] has become increasingly influenced and determined by the presence of spatially immobile clusters of value-added activities (Dunning, 1997). In effect, the production is to a large extent locally bounded. Proximity and location play an essential role in the transmission of knowledge. This stems from the tacit elements of knowledge and from the interactive nature of the innovation process.

It is for this reason that the optimal level of governance is requiring re-examination. Cooke and Morgan (1998) and several others claim that it is the regional level that offers the greatest prospects for devising governance structures which enable to foster the transition towards a learning economy. Thus, it has been frequently emphasised – despite continued predictions of the 'end of geography' – that regions remain a crucial component in the organisation of economic activity. Florida (1995, p. 527), for example, argues that 'regions are becoming focal points for knowledge creation and learning, as they in effect become learning regions. These learning regions function as collectors and repositories of knowledge and ideas, and provide the underlying environment or infrastructure which facilitates the flow of knowledge, ideas and learning'. This argument is based on the observation that knowledge creation is context-dependent or, in other words, embedded in socio-economic networks which rely on close interaction and exchange of tacit knowledge. Learning regions are not only important sources of innovation and economic growth, but also – via their globe-straddling [producer/consumer-driven] networks – vehicles for globalisation.

Despite several outstanding studies of the dynamics of individual regions (see, Saxenian, 1994, for example), the role of regions in the new age of market-based capitalism remains still rather poorly understood. Even though the contributions included into the present book do not attempt to develop a theory of what such changes imply for regions, regional organisation and regional development, at least some of them shed some further light on this issue that is of central importance for our discipline.

1.2 The Structure of the Book

The book has been divided into four parts which group the chapters according to their main focus. PART A looks at innovation systems, learning regions and human resources. The notion of a learning region is at the core of an increasing number of regional development strategies. The notion is largely based on the systems of innovation approach. Though relatively new, it is at the core of an increasing number of regional development strategies. It is the innovative capacity of organisations [predominantly firms] that to a large extent determines their competitiveness within the learning region. PART B shifts our attention to the issue of competitiveness. The term competitiveness, applied to regions, is more than the simple sum or the average of the competitiveness of the firms which operate within their boundaries. Particular emphasis will be laid on the role of public institutions for regional competition, but also on the issue of cooperation. PART C directs the focus of attention to the role of transport infrastructure in economic development in the specific context of the European Union [EU]. Different ways of modelling and measuring this relationship, and the development of policy are considered, enriched with four energy scenarios for Western European passenger transport in the twenty-first century. PART D concludes the book with a special focus on the opportunities and threats for regional development and policy in face of the increasing integration that is taking place in Europe in these days.

PART A: Innovation Systems, Learning Regions and Human Resources

While there is considerable debate on the precise nature of the restructuring there is a widespread agreement that knowledge creation, and diffusion, and learning are significant processes in determining economic development and competitiveness. It is clear that these processes of economic transformations have impacts which are differentiated between different localities. Some regions are better able to adapt to the new economic situation than others. The distinctive characteristics of regions – their economic structures, patterns of social and political relations and their social capital in the form of business culture – are themselves critical in shaping future patterns of development (Edquist and Rees, 2000).

Florida (1995) argues that the new age of capitalism requires a new kind of region, regions that adopt to the principles of knowledge creation and continuous learning. Regions must in effect become learning regions. There is no single definition of a learning region, but the notion draws very much on the systems of innovation approach (see, for example, Lundvall, 1992; Edquist, 1997; Braczyk et al., 1998; Fischer, 2001; Fischer and Fröhlich, 2001).

The systems of innovation approach carries the idea that innovations do not originate as isolated discrete phenomena, but are generated by means of the interaction of a number of actors or agents. These sets of actors and interactions have certain characteristics that tend to persist over time. The same characteristics are shared by national, regional, sectoral and technological innovation systems. They can all be represented as sets of actors and interactions whose ultimate goal is the production and diffusion of knowledge. A major theoretical and policy problem posed by innovation systems is that knowledge is generated not only by individuals and organisations [firms, educational organisations, R&D institutions, departments and agencies of the national/regional state, intermediate organisations], but also by their often complex patterns of interaction. Viewed from this perspective a learning region may be seen as a regional innovation system or as a model towards which actual regions need to develop to meet most effectively the challenges posed by the transition towards a learning economy.

The first chapter in PART A, by *Manfred M. Fischer*, makes a modest attempt to increase our ability to understand the systems of innovation approach as a flexible and useful conceptual framework for innovation analysis. It presents an effort to develop some missing links and to decrease the conceptual noise often present in the discussions on spatial innovation systems. The chapter specifies elements and relations that seem to be essential to the conceptual core of the framework and argues that there is no a priori reason to emphasise the national over the subnational [regional] scale as an appropriate mode for analysis, irrespective of time and place. Localised input-output relations between the actors of the system, knowledge spillovers and their untraded interdependencies lie at the centre of the arguments. The chapter is organised as follows. It introduces the reader, first, to some basic elements and concepts that are central to understanding the approach. The characteristics of the innovation process are examined: its nature, sources and some of the factors shaping its development. Particular emphasis is laid on the role of knowledge creation and dissemination based on the fundamental distinction between codified and tacit forms. These concepts recur throughout this and other chapters and particularly in discussions on the nature and specifications of the systems approach. The chapter concludes by summarising some of the major findings of the discussion and pointing to some directions for future research activities.

The interactive nature of innovation, the blurring of the distinction between production and diffusion of technology, and the adoption of network relationships by and among firms places the enterprise at the centre of contemporary changes. The second chapter, written by *Riccardo Cappellin*, discusses the implications of these changes for the corporate organisation on the one side and for regional development and policies on the other.

The new economic realities make it indispensable to competitiveness that human resources adapt to technological change and contribute to shaping and diffusing technology with rapidly changing technologies. This requires not only appropriate basic training and skills, but also long-life education and retraining. The last chapter in PART A, written by *Peter Batey*, addresses issues of the relationship between human resources and regional development. The chapter presents a survey of current thinking about the human resources component of regional development strategies. It begins by drawing upon a sample of international evidence of variations in education and training levels across the range of OECD countries. This international comparison provides a backcloth to the regional comparison that follows in later sections of this chapter. The author continues to review a number of key concepts, especially the notions of a learning economy and a learning society and the subsequent translation of these ideas into the concept of a learning region. The chapter goes on to discuss the special role of universities and to consider indicators and targets that can be utilised to chart the progress of human resource development in a system of regions. It concludes by examining two examples of regional human resource development strategies. The examples show the need to stimulate demand for work-related education and training and the steps that must be taken to overcome social exclusion. Increasing access to learning opportunities has the effect of improving the people's employability and, hence, reducing social exclusion. The second of the two examples also demonstrates the linkages between different tiers of human development strategy, at European Union, national and regional levels.

Part B: Competitiveness and Cooperation

In the current policy debate the notion of competitiveness plays a central role. But the notion, while relatively obvious when applied to firms, is less clear when applied to regions. Transferring the notion of competitiveness from the sphere of firms to the mesoeconomic level, that of regions, is marked by considerable difficulties and intellectual dangers. The analogy between the competitiveness of a firm and the competitiveness of a region – often the basis of popular theorising – is week indeed (OECD, 1992).

Competitiveness of a region is built on the competitiveness of the firms operating within its boundaries. It is – to a large extent – an expression of their ability to sell their products and services on national and international markets, to increase their profits and to grow. Successful competition requires quality of products as much as inherent novelty, superior process technology and organisation of production, speed of delivery and quality after-sales services. The competitiveness of firms today is increasingly shaped by the various aspects of corporate organisation including the capacity to incorporate in-house R&D and innovation-related activities with R&D cooperation with universities and other organisations, the capacity to organise successfully inter-firm relationships with component and material supplier firms upstream and with retailers downstream, and the steps taken to enhance worker's skills through investments in training (OECD, 1992). Thus, the competitiveness of firms obviously reflects successful management practice. It also stems from region-specific long-term trends in the strength and efficiency of a regional economy's productive structure, its technical infrastructure and other externalities on which firms build.

The competitiveness of a region is evidently more than the simple sum or the average of the competitiveness of its firms. The notion has to be based upon a more holistic approach to include dimensions such as the regions's ability to generate high income and employment levels and to reflect institutional factors that embrace a wide range of phenomena which relate to the way regional economies hold together and work.

An unfavourable sectoral structure together with a lack of innovative capacity (Suarez-Villa, 2000) seems to be among the most important factors underlying lagging competitiveness. In contrast, successful regions tend to offer competitive advantages that fall under four broad headings, essentially those of Porter's (1990) competitive diamond:

- *First,* supply of infrastructure in terms of the provision of information, communication and research infrastructures, and educational opportunities, but also the quality of available labour and the availability of capital.
- *Second,* demand in terms of the propensity of regional customers to become players in the regional innovation process through their demands and their capacity for initiative and innovation.

- *Third,* quality of the regional industrial fabric [in terms of subcontractors, suppliers of intermediate goods and equipment and producer services suppliers] and the social capital [in terms of the norms, values and other social conventions that are shared in every day interactions within a social network], facilitate exchange of knowledge, and reflect specific paths or trajectories of economic, social or cultural development.
- *Fourth,* organisation and managerial practices of firms as well as efficiency of public institutions.

These four dimensions reinforce each other, lack of any of these dimensions will jeopandise the region's potential for development (OECD, 1992).

The first contribution to PART B, written by *Peter Friedrich* and *Xiao Feng,* attempts to shed some light on the role of public institutions for regional competitiveness in general, and regional competition by public institutions in particular. Special focus is laid on three issues in this context: Specification issues [that is, specification of regional competition among public institutions], modelling issues [that is, how to model political public regional competition] and measurement issues [that is, how to measure competitiveness with special reference to the role of public institutions].

There may be competition between regions, leading to winners and losers, but there may also be cooperation yielding joint growth and cooperation (Corvers and Nijkamp, 2000). The chapter, by *Manas Chatterji* and *Bengt Lorendahl*, shifts attention to the issue of regional cooperation in the context of the Baltic States. The authors stress the point that cooperation between countries and regions within a country is essential for conflict management and peaceful coexistence. The objective of this contribution is to develop some comparative perspectives to characterise these resource based conflicts and to demonstrate how different models of cooperation can be used to devise policies to strengthen the historical bands of the Baltic States. The authors claim that the discussion though focussed on Baltic States is also relevant to characterise the relationship between the territory of the United States and Canada which borders the Great Lakes and elsewhere.

PART C: Mobility and Transport Infrastructure

A fundamental prerequisite of the evolution of the global, knowledge-based age of capitalism is the development of technologies that overcome the 'tyranny of space and time'. The most important of such technologies are the IT-based technologies of transport and communications. Neither of these technologies can be considered as the cause of the current shift to the new economy. But they provide the indispensable material basis for this economy as Castells (1996) emphasises.

The chapter, written by *Roger Vickerman*, directs our attention to the role of transport infrastructure in economic development in the specific context of the European Union. The author considers ways of modelling and measuring this relationship, and the development of policy. A primary focus is on the problem of combining policies that seek to improve the competitiveness of the EU, and those that seek to increase cohesion and promote convergence of GDP per capita, especially in the context of expansion of the EU to include Central and Eastern European countries. The chapter concludes in drawing attention to three key problems: *First*, the author identifies a conflict between policies towards achieving internal consistency in the transport sector and the role of transport in the wider economy. *Second*, he points to a potential conflict between competitiveness and cohesion. *Third*, there has been a possible excess emphasis on large projects of EU interest to the exclusion of local projects that could have had wider significance.

The discussion in this chapter shows that our understanding of the ways in which transport infrastructure affects economic development and the process of integration is still rather limited. The major problem we face is that there is an assumption that infrastructure must have a positive effect on growth, and, thus, on competitiveness and cohesion. In the next chapter, *Roger Vickerman* continues to review various theoretical approaches that may be utilised to assess the wider and longer-term impacts of transport infrastructure investment. It is shown that this is a complex process which demands more than the estimation of a single aggregate relationship between infrastructure investment and the level of economic activity. The key elements required are a sufficient representation of both the sectoral and spatial structures of the relevant economy to capture all of the changes which the investment may induce. The key elements are an appreciation of the competitive structure of the markets affected, the significance of transport costs in both the provision of inputs and the delivery of outputs for that sector as well as the scope for substitution of transport for other inputs.

Predicting the future in transport is generally fraught with major difficulties and serious intellectual dangers in general and especially in the context set by the contemporary transition to a global knowledge driven economy and society. *Peter Nijkamp* and associates develop – on the basis of eight observable phenomena [so-called 'force-fields'] – four energy scenarios for Western European passenger transport in the twenty-first century. The present transport system serves as baseline case. For each scenario the authors sketch how the passenger transport system may look like in terms of the use of various existing and future transport technologies and the corresponding modal split. Based on this information, expected energy consumption features of the various transport modes are described, data on the present fuel supply and electricity generation system are offered, and estimations of the future energy system are given. Then the energy consumption and CO_2 emissions associated with the future passenger transport systems are assessed and these impacts compared with the current system. The authors make the point that these scenarios provide interesting policy options for decision makers. There is no doubt in mind that an environmentally friendly transport system is feasible but at the expense of drastic policy measures which may affect economic growth and the lifestyles of individuals.

PART D: Regional Development and Policy

The emergence of the global knowledge-driven economy, with all its uncertainties and the difficulties of national macro-economic management, demonstrates that national and regional governments need to reconsider their role and how their actions may affect the competitive advantages of their regions and of the firms operating within their boundaries. While globalisation will require governments to surrender some of their traditional tasks to market forces, those which remain are no less vital to the economic welfare. On the one hand, an increasing role is being recognised for subnational [regional] levels of governance (see Braczyk et al., 1998) and, on the other, new or strengthened supranational governance institutions may be necessary.

The European Union – obliged by the Single European Act to reduce the differences between regions to realise the objective of social cohesion – has taken on an important role in regional policy throughout Europe in the recent past. Through the European Regional Development Fund that provides regional subsidies throughout the EU, the supranational organisation approves nationally designated regions for assistance. Objective 1 [regions lagging behind in development, measured in terms of GRP], Objective 2 [areas in industrial decline], Objective 5b [vulnerable rural areas] and Objective 6 [regions with very low population density] are considered to represent EU regional policy.

The chapter, written by *Gündüz Atalik* [a reprint from the *Papers of the Regional Science Association*, Vol. 69, 11-19] aims to discuss integration problems concerning regional differentiation and regional disparities in the EU, by adapting theoretical approaches and indicators for measuring the disparities in regional development policies. To test the relevance of the potentiality factors, the analysis studies the causes of regional disparities that are used in the studies as the most convenient approach. Turkish GDP variations by regions are analysed using the quasi-production function based on a Cobb-Douglas specification taking into account as a function of activity, infrastructure, industrial employment and agglomeration rates. The results show that these factors are significant for testing the potential theory for a developing area.

In face of the contemporary transition to the new stage of market-based capitalism not only supranational organisations like the European Commission, but also national and regional governments will have to reconsider their role. In doing so governments should become aware of the need to envisage their efficiency in the broader framework of an overall policy strategy to facilitate flows of knowledge and other interactions between the different components of an innovation system. Hereby the following trends and principles should be taken into account (Fischer et al., 2001):

- *First,* there is an overall shift from traditional supply-side diffusion measures towards policies, which reflect a more interactive model of innovation and recognise innovation and diffusion as interdependent processes. This translates into emphasising a greater role for demand-driven policy programmes, network-building initiatives, and policy measures to upgrade the technology diffusion infrastructure and improve its relevance for and accessibility by smaller businesses.
- *Second,* the nature of knowledge being diffused is changing. Greater emphasis has to be laid on diffusing soft organisational technologies of both technical and quality management as well as on hard technology of information and communication management.
- *Third,* severe constraints on public finances have robbed governments of the means even if they still have the will to engage in large-scale and widespread interventions. Like firms governments face a difficult learning process nowadays as they work out to pursue the public interest efficiently and effectively in an increasingly global and increasingly knowledge-based economy. Governments cannot leave entirely to the discretion of firms and other organisations the choice of organisational adjustment processes in the face of globalisation and rapid technological change. They have to remain powerful framers of the business environment even in a more global knowledge-based economy. They will have to make choices between market-reliant, market-regulating, market-supplementing and market-displacing strategies. A redesign of the environment for business is necessary as foreign direct investment and the internet economy become more and more important features of the global economy, and as the knowledge intensity of internationally traded goods and services increases.
- *Fourth,* there is a need for policy coordination, to rank order the main policy objectives, build the overall innovative capacity of firms, and stimulate the adoption of specific technologies. This is crucial for achieving greater coherence between programmes with different targets [specific technologies, sectors, institutions or regions] and offering different diffusion services as technical assistance, information and training.
- *Fifth,* in order to arrive at maximum leverage, technology policies should build on existing relationships in innovation systems, in particular networks within which firms collaborate and exchange tacit knowledge.

The major challenge for local and regional policy lies in the development of institutions which are able to foster restructuring and to stimulate regional networks, but also to link up the regional economy with international and global firms, markets and networks. The last contribution to PART D, by *Daniela L. Constantin,* addresses this issue in a post-socialist context where institutional reforms are at least as important as macro-economic reforms to create an environment for encouraging competition and sustained economic growth. In most Central and East European countries ministries responsible for regional development play essentially a policy development role and, in some cases such as Romania a coordinating rather than an operational role. They prepare draft legal acts and regional policy reform. The author does not only direct our attention to the regional policy reform in Romania but also to the challenges Romania is facing with the introduction of a regional policy dimension.

Certainly, the contributions presented in this volume do not tell the whole story of the rediscovery of the region as a fundamental basis of economic and social life, but they can be considered an attempt to throw light on the complexity of regional development issues, and hopefully provide useful indications for both further analysis and future policy.

References

Archibugi D. (2001): The Globalisation of Technology and the European Innovation System. In: Fischer M.M. and Fröhlich J. (eds.) *Knowledge, Complexity and Innovation Systems*, Springer, Berlin, Heidelberg, New York, pp. 58-75

Braczyk H.-J., Cooke P. and Heidenreich M. (eds.) (1998): *Regional Innovation Systems. The Role of Governances in a Globalized World*, UCL Press, London

Castells M. (1996): *The Rise of the Network Society*, Blackwell, Oxford

Cooke P. and Morgan K. (1998): *The Associational Economy*, Oxford University Press, Oxford

Corvers F. and Nijkamp P. (2000): Regional Development and Interorganisational Policy Making in a Pan-European Context. In: Kohno H, Nijkamp P. and Poot J. (eds.) *Regional Cohesion and Competition in the Age of Globalisation*, Edward Elgar, Cheltenham, pp. 265-285

Daniels P.W. and Lever W.F. (eds.) (1996): *The Global Economy in Transition,* Addison Wesley, Longman, Essex

Dicken P. (1998): *Global Shift. Transforming the World Economy*, The Guilford Press, New York

Drucker P.F. (1993): *Post-Capitalist Society,* Butterworth Heinemann, Oxford

Dunning J.H. (1997): Technology and the Changing Boundaries of Firms and Governments. In: OECD (ed.) *Industrial Competitiveness in the Knowledge-Based Economy: The New Role of Governments*, Organisation for Economic Cooperation and Development, Paris, pp. 53-67

Edquist C. (ed.) (1997): *Systems of Innovation. Technologies, Institutions and Organisations*, Pinter, London

Edquist C. and Rees G. (2000): Learning Regions and Cities: Learning in Regional Innovation Systems – A Conceptual Framework, Paper Presented at the International Workshop on 'Knowledge, Complexity and Innovation Systems', Vienna, July 1-3, 2000

Fischer M.M. (2001): Innovation, Knowledge Creation and Systems of Innovation, *The Annals of Regional Science* 35 (2), 199-216

Fischer M.M. and Fröhlich J. (eds.) (2001): *Knowledge, Complexity and Innovation Systems*, Springer, Berlin, Heidelberg, New York

Fischer M.M., Revilla Diez J. and Snickars F. (2001): *Metropolitan Innovation Systems. Theory and Evidence from Three Metropolitan Regions in Europe*, Springer, Berlin, Heidelberg, New York

Florida R. (1995): Towards the Learning Region, *Futures* 27 (5), 527-536

Guinet J. (1997): Knowledge Flows in National Innovation Systems. In: OECD (ed.) *Industrial Competitiveness in the Knowledge-Based Economy,* Organisation for Economic Cooperation and Development, Paris, pp. 173-178

Lundvall B.-Å. (ed.) (1992): *National Systems of Innovation: Towards a Theory of Innovation and Interactive Learning*, Pinter, London

Malecki E.J. (1997): *Technology & Economic Development. The Dynamics of Local, Regional and National Competitiveness,* Addison Wesley, Longman, Essex

Morgan K. (1997): The Learning Region: Institutions, Innovation and Regional Renewal, *Regional Studies* 31 (5), 491-503

Nonaka I. and Takeuchi H. (1995): *The Knowledge-Creating Company. How Japanese Companies Create the Dynamics of Innovation*, Oxford University Press, Oxford, New York

OECD (1992): *Technology and the Economy: The Key Relationships,* Organisation for Economic Cooperation and Development, Paris

OECD (1996): *Transitions to Learning Economies and Societies*, Organisation for Economic Cooperation and Development, Paris

Porter M.E. (1990): *The Competitive Advantage of Nations*, MacMillan, London, Basingstoke

Saxenian A. (1994): *Regional Advantage: Culture and Competition in Silicon Valley and Route 128,* Harvard University Press, Cambridge [MA]

Storper M. (1997): *The Regional World. Territorial Development in a Global World*, The Guilford Press, New York, London

Suarez-Villa L. (2000): *Invention and the Rise of Technocapitalism*, Rowman & Littlefield Publishers, Lanham, Boulder, New York, Oxford

PART A: Innovation Systems, Learning Regions and Human Resources

2 A Systemic Approach to Innovation

Manfred M. Fischer
Vienna University of Economics and Business Administration
Department of Economic Geography & Geoinformatics

2.1 Introduction

The context of this chapter is set by the contemporary transition from an industrial to a knowledge-based economy. OECD (1996a) has documented the way in which the knowledge-based economy of learning individuals, organisations and economies is emerging out of the machine-based economy that determined advanced economies in most of the time in the 20th century. Although there is considerable debate on the precise nature of the restructuring that is taking place, there is wide agreement that knowledge creation and diffusion are increasingly significant processes in the innovation process. The innovation process involves the use, application and transformation of scientific and technological knowledge in the solution of practical problems.

Such an innovation process requires a high level of interaction, dialogue and exchange of information that may be conducted long-distance, but is often less expensive, more reliable and easier to conduct locally. Recent empirical studies support this view (see, for example, Jaffe et al., 1993; Anselin et al., 1997; Fischer and Varga, 2001). The clustering of inputs such as industrial R&D and university research, agglomerations of manufacturing firms in related industries and networks of producer service providers often create scale economies in the generation of knowledge, and facilitate the transfer of knowledge to firms in the region (Maskell, 1999).

The emphasis on knowledge creation and interactivity provides a foundation for systemic approaches to the analysis of innovation processes. Systems of innovation is a novel systemic approach that has emerged more or less during the past decade for studying innovation as an endogeneous part of the economy and, in fact, as driving force behind economic change. According to this approach innovations are seen as part of a larger process of knowledge production of economic relevance. The approach stresses that firms do not innovate in isolation, but in interaction with other organisations. These may include other firms [suppliers, customers and competitors], R&D institutions, universities and other forms of producer services.

In this chapter, we provide a more precise view of the innovation process first, then address the role of knowledge creation and spillovers, present the systems of innovation approach, make a strong case for the importance of the subnational regional scale as an appropriate mode of analysing, and conclude finally with a brief discussion of policy implications of this perspective. The objective is to increase our ability to understand the importance of knowledge creation and innovation for economic performance.

2.2 The Innovation Process

Commercial products and production processes represent various combinations of pieces of knowledge in a specific technology set. Innovation is generally defined as the activities of developing and commercialising new products and processes (see, for example, Hall, 1986). These innovation activities are of two major types: fundamental, which involves the creation and utilisation of a piece of novel scientific, technological or organisational knowledge; and incremental, which concerns product or process improvements based on existing knowledge (Freeman, 1986). Product innovations are a matter of *what* is produced, while process innovations are a matter of *how* products are produced. Technological products and process innovations in the form of goods yield material outcomes. Organisational process innovations and product innovations in services are intangible. But they are not less important (Edquist and Rees, 2000).

For a long time thinking about innovation has been guided by a linear model. The development, production and marketing of new technology [defined in the sense of Mansfield et al. (1982) as consisting of a pool or set of knowledge] was assumed to follow a well defined time sequence which began with basic and applied research activities, involved a product development stage, and then led to production and possibly commercialisation. This model corresponds less and less to the realities of the innovation process. It was undoubtedly better adapted to the technology trajectories based on radical process innovations, as in the 1950s and 1960s [for example, petrochemicals development] than today's that are more based on product innovation and even, to some extent, eliminate the distinction between these two types of innovation.

This criticism has led to a broader view of the process of innovation as an interactive process. The presently emerging innovation theory emphasises the central role of feedback effects between the downstream and upstream phases of innovation and the numerous interactions between science, technology and innovation related activities within and among firms. Through interaction and feedback different pieces of knowledge become combined in new ways or new knowledge is created.

The innovation process at the firm level may be viewed as a set of activities that are linked to one another through complex feedback loops. The process may be seen as a chain, starting with the perception of a new market opportunity and/or a new invention based on novel pieces of scientific and/or technological knowledge followed by the analytical design for a new product or process and testing, redesign and production, and distribution and marketing. Short feedback loops link each downstream phase with the phase immediately preceding it. Longer feedback loops link perceived market demand and product users with phases upstream. A second set of relationships link the innovation process embedded in the firm with its firm-specific knowledge base, the general scientific and technological pool and with research in general (see Fischer, 2001).

This so-called chain-linked model of the innovation process (see Kline and Rosenberg, 1986) combines two types of interaction. One concerns processes that occur through new forms of product development practice within the firm and create appropriate feedback relationships (see, for example, Nonaka and Takeuchi, 1995). The second refers to relationships external to a given firm with other companies such as customers, suppliers of inputs [including finance and knowledge], research institutions and even competitors. Cooperation can take place with various mixes of internal and external actors. Under this model technological innovation is seen to be the result of a complex interplay among various actors, with partly common and partly conflicting interests. Technological progress is, thus, dependent on how the actors interact with each other, internally and externally.

In recent years, new forms of interfirm agreements bearing on technology have developed alongside the traditional means of technology transfer – licensing and trade in patents – and they often have become the most important way for firms, regions and nation-states to gain access to new knowledge and key technologies. The network form of governance – involving both action by mediating third parties as well as 'relational contracting' between firms – can overcome market imperfections as well as the rigidities of vertically integrated hierarchies. The limitations of these two modes of transaction in the context of knowledge and innovation diffusion have pushed interfirm agreements to the forefront of corporate strategy in the last few decades (Chesnais, 1988).

There are many definitions of *innovation networks* (see De Bresson and Amesse, 1991; Freeman, 1991); however, the one offered by Tijssen (1998, p. 792) captures the most important features of the network mode. He suggests defining a network as '*an evolving mutual dependency system based on resource relationships in which their systemic character is the outcome of interactions, processes, procedures and institutionalisation. Activities within such a network involve the creation, combination, exchange, transformation, absorption and exploitation of resources within a wide range of formal and informal relationships.*' In a network mode of resource allocation, transactions occur neither through discrete exchanges nor by administrative fiat, but through networks of individuals or institutions, engaged in reciprocal, preferential and supportive actions (Powell, 1990).

Networks show a considerable range and variety in content, which differs according to specific circumstances. Their nature is shaped by the objectives for which network linkages are formed. For example, they may focus on a single point of the R&D-to-commercialisation process or may cover the whole innovation process. The content and shape of a network will also differ according to the nature of relationships and linkages between the various actors involved (see Chesnais, 1988). At the one end of the spectrum lie highly formalised relationships. The formal structure may consist of regulations, contracts and rules that link actors and activities with varying degrees of constraint. At the other end are the network relations of a mainly informal nature, linking actors through open chains. Such relations are very hard to measure (Freeman, 1991). Whenever interfirm transactions tend to be small in scale, variable and unpredictable in nature, requiring face-to-face contact, then network formation will focus on the close proximity of the partners involved (Storper, 1997). For a review that attempts to connect the formation of networks with the knowledge-based economy, see Fischer (2002).

2.3 Knowledge Creation and Spillovers

The system of innovation approach places knowledge creation and knowledge spillovers at its very centre. Any discussion of knowledge invariably leads to the question of the relationship between information and knowledge. The common understanding is that information does not become knowledge unless its value is somehow enhanced through interpretation, organisation, filtration, selection or engineering. Information may be interpreted as factual (Saviotti, 1988), while knowledge establishes generalisations and correlations between variables (Andersson, 1985). Particular pieces of information can be understood only in the context of a given type of knowledge, for example a theory. New knowledge creates new information and this information can be understood and used only by those who possess the new knowledge. In this sense knowledge has a *retrieval/interpretative* and not only a *correlational* function (Saviotti, 1998).

Knowledge has some further outstanding characteristics that are important to mention. *First,* knowledge is cumulative (Teece, 1981; Nelson and Winter, 1982) which implies path-dependence and the possibility of creating barriers, since established participants – in given technologies – accumulate a differential advantage with respect to potential entrants. Knowledge in firms also has a collective character. This means that knowledge is not simply the sum of the pieces embodied in the individual workers of the firm (Saviotti, 1998). In this sense, the knowledge base of a firm may be defined as the collective knowledge that a firm uses to produce its output.

The knowledge base contains knowledge in all its forms, from simple and routine procedures of everyday life to the methods of organisation and management, from the machinery [i.e. embodied knowledge] to the scientific concepts, methods and theories that enable newer inventions. In most cases, a piece of knowledge can be located somewhere in a range between the completely tacit and completely codified. Tacitness refers to those elements of knowledge that persons have which are ill-defined, uncodified and which they themselves can not fully articulate and which differ from person to person, but which may be shared by collaborators who have a common experience (Polanyi, 1966). Knowledge is always at least partly tacit in the minds of those who create it. The process of codification is necessary because knowledge production is a collective undertaking that requires communication. The transmitter and the receiver have to know the code to be able to communicate. The codification process for a given subject amounts to the gradual convergence of the scientific community and of other users on common standardised definitions and concepts, on common contents and theories. The degree of codification differs for different types of knowledge at a given time. Knowledge closer to the frontier, and thus more recent, is likely to be more tacit than knowledge which is already established (Saviotti, 1998).

Codified knowledge is that form of knowledge that is in some way tangible – usually found in print form, such as scientific papers and patent applications. Much knowledge is codified and accessible. But much of the essential knowledge – especially the newer parts that constitute the frontier – resides in tacit form in the minds of experienced individual researchers or engineers. This person-embodied knowledge is generally difficult to transfer, and is often only shared by colleagues if they have learned the code through common practice. On the one hand, a given type of knowledge may become more codified as it matures, on the other, the act of embodying it into specific goods and services may reintroduce some 'tacitness'.

Spillovers stem from specific features of knowledge. In particular, knowledge is a non-rivalrous and partially excludable good. Non-rivalry implies that a new piece of knowledge can be utilised many times and in many different circumstances, for example by combining with knowledge coming from another domain. Lack of excludability, on the other hand, implies that it is difficult for firms that have devoted resources to R&D fully to appropriate the benefits and prevent others from using the knowledge without compensation or with compensation less than the value of the knowledge (Teece, 1986). While knowledge is subject to spillovers, however, it is only imperfectly excludable. With the use of patents or other devices such as secrecy knowledge producing firms capture at least part of the social benefits associated with the production of knowledge, and this is an incentive for their R&D investment (OECD, 1992). The interest of users of knowledge [i.e. firms other than the knowledge producing firm] is thus best served if – once produced – knowledge is widely available and diffused at the lowest possible cost. This implies low appropriability for knowledge producers or – put in another way – an environment rich in knowledge spillovers.

The term *spillover* is used in economics to capture the idea that some of the economic benefits of R&D activities accrue to economic agents other than the party that undertakes the research. Competing firms that initiate a successful innovation, and firms whose own research benefits from observation of the successes and failures of others' research effects all garner such spillover benefits. These examples suggest that such spillovers are created by a combination of the new knowledge resulting from a R&D effect, and the commercialisation of the new technology in terms of a new product or process that is successfully implemented in the market place (Jaffe, 1996). Research spillovers have been defined by Cohen and Levinthal (1989) to include any original valuable knowledge generated in the research process that becomes publicly accessible whether it be knowledge fully characterising an innovation or knowledge of a more intermediate nature. They have been also termed disembodied or *knowledge spillovers* to emphasise that they do not necessarily relate to knowledge embodied in machinery or equipment.

2.4 The Systems of Innovation Approach

The systems of innovation approach places innovation, knowledge creation and diffusion at its very centre. Innovation and knowledge creation are viewed as interactive and cumulative processes contingent on the institutional set-up. It departs from the network school of research (Håkansson, 1987) with its emphasis on the role of the institutional set-up, i.e. that institutions play in the innovation process (see Edquist and Johnson, 1997). The concept of institutions refers at an abstract level to the recurrent patterns of behaviour: socially inherited habits, conventions including regulation, values and routines (Morgan, 1997) that assist in regulating the relations between people and groups of people within as well as between and outside the organisations. Where the interrelationships between organisations are characterised by high levels of trust, uncertainty in relation to knowledge exchange is reduced, stable and reciprocal interactions are developed and, thus, innovative capability is greatly enhanced (Edquist and Rees, 2000).

A *system of innovation* can be thought of as comprising of a set of actors or entities such as firms, other organisations and institutions that interact in the generation, use and diffusion of new – and economically useful – knowledge in the production process. At the current stage of development there is no general agreement as to which elements and relations are essential to the conceptual core of the framework and what is their precise content (Edquist, 1997b). This leaves room for a conceptual discussion.

Systems that attempt to encompass the whole innovation process may be expected to include four key building blocks that consist of groups of actors sharing some common characteristics and institutions governing the relations within and between the groups (see Fig. 2.1):

- *The Manufacturing Sector*
  This sector is made up of manufacturing firms [the central actors in the system of innovation] and their R&D laboratories that play a fundamental role in performing research and technological development.
- *The Scientific Sector*
  The scientific sector plays a very important role in technological innovation. It consists of two components: a training component that includes educational and training organisations on which the supply of scientists, engineers, technicians and other skilled workers possessing appropriate skill profiles depends, and a research component including universities and other research organisations that generate and diffuse knowledge and produce documents in the form of scientific publications. This sector involves those agents [government, private non-profit, universities, higher education] that both fund and carry out research or offer education.
- *The Sector of Producer Services*
  This sector includes organisations or units within larger organisations which provide assistance or support to industrial firms for the development and/or introduction of new products or processes. This may take any of the following forms: financial, technical advice or expertise, physical [equipment, software, computing facilities], marketing or training related to new technologies or procedures.
- *The Institutional Sector*
  Many of the tasks that a typical firm must perform require coordination, either within the firm between various groups of employees or outside it with other suppliers, other firms, and providers of producer services including finance.

There is a variety of ways in which the performance of these tasks can be coordinated, each involving different kinds of firm behaviour. But in general one can distinguish market coordination that relies on the kind of market institutions neo-classical economics usually assumes to be important, and non-market coordination that utilises a greater range of institutional arrangements. The latter depends upon the presence of formal and informal institutions that regulate the relations between the actors of the system, enhance their innovation capacities and manage conflicts and cooperation. Two types of such institutions may be distinguished (see, for example, Edquist and Johnson, 1997): formal institutions including, e.g., employer associations, legal and regulatory frameworks, and informal institutions including the prevailing set of rules, conventions and norms that prescribe behavioural roles and shape expectations.

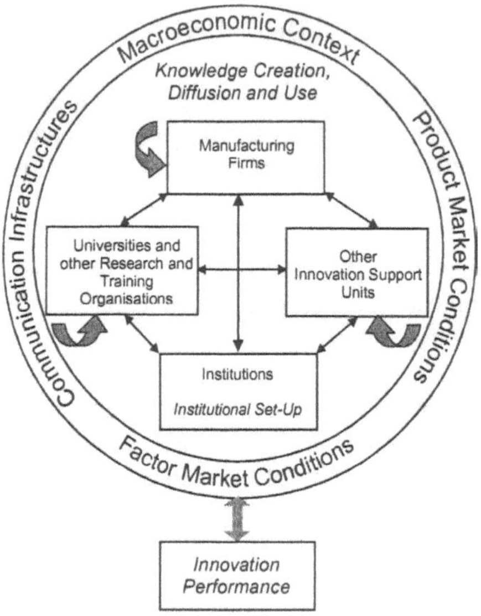

Fig. 2.1 The major building blocks of a System of Innovation

To describe and compare systems of innovation in the broad sense one has to open the boxes of the subsystems, identify the constituent elements and specify those relations between and within the subsystems that have importance for innovation performance. A first source of diversity among different systems might be due to differences in the macroeconomic context, the quality of information and communication infrastructures as well as in factor and product market conditions.

The innovation performance of an economy is notably determined by the characteristics and abilities of its individual firms and other organisations contingent on its institutions, but also very much by the different kinds of relations between them, i.e. the ways they interact with each other and with the sector of institutions. The character as well as the change of these interaction patterns are central aspects of innovation systems. Linkages within and between the sectors can be specified in terms of knowledge and information flows, flows of investment funding, flows of authority and labour mobility [scientists, technicians, engineers and other skilled workers] as important mechanisms for the transfer of tacit forms of knowledge particularly from the scientific to the manufacturing sector, but also within the latter.

Network analysis may assist to identify the central actors in the four subsystems [building blocks] in specific cases, and of the type of information and knowledge they exchange. Different kinds of norms, conventions and established practices that are expected to have important implications for knowledge creation and learning are forming the economy's patterns of interaction, both inside firms and other organisations and between them. Searching for and explaining interaction patterns that lead to the creation, dissemination and use of knowledge is part of the systems of innovation approach (Johnson, 1997). It can be hypothesised, for example, that there will be strong and weak, regular and irregular interactions which shape the system.

Firms are the main carriers of technological innovation. Their capacity to innovate is partly determined by their own capabilities, and partly by their absorption capacities. Increasing complexity, costs and risks in innovation enhance the role of collaboration and networking in the innovation process to reduce moral hazard and transaction costs. In addition to traditional market-mediated relations such as the purchase of equipment and licensing of technology, firms exchange information and engage in mutual learning in their roles as customers, suppliers and subcontractors, and even competitors. A coherent system of innovation has necessarily to include a series of more or less coordinated network-like relations (Fischer, 1999) such as:

- *Customer-producer relations,* i.e., forward linkages of manufacturing firms with distributors, value-added resellers and end users,
- *Producer-manufacturing supplier relations* which include subcontracting arrangements between a client and its manufacturing suppliers of intermediate production units,
- *Producer-service supplier relations* which include arrangements between a client and its producer service partners [especially computer and related service firms, technical consultants, business and management consultants],
- *Producer network relations* which include all co-production arrangements [bearing on some degree or another on technology] that enable competing producers to pool their production capacities, financial and human resources in order to broaden their product portfolios and geographic coverage,
- *Science-industry collaboration* between universities and industrial firms at various levels pursued to gain rapid access to new scientific and technological knowledge and to benefit from economies of scale in joint R&D, such as direct interactions between particular firms and particular faculty members, or joint research projects, as through consulting arrangements, or mechanisms that tie university or research programs to groups of firms.

2.5 Regional Systems of Innovation

Interest in regional innovation systems arises from the manifest differences in economic growth and levels of welfare between regions. Regions within nation-states share some of the aspects of the entire countries such as educational legislation and technology policy, but they have different possibilities to go their own way, and ultimately end up diverging from a national average. Especially if there is a large number of firms that form a specialised local/regional system, it is likely that the region becomes distinct from the rest of the regions in a country in some significant aspects (Oinas and Malecki, 1999).

Geographic proximity can be considered as a *necessary, but not sufficient* precondition for the existence of a regional system of innovation. A proximity that is only geographic in nature can provide the basis for the presence of an agglomeration of firms, but not necessarily for the presence of a system of innovation. The potential of an innovation system crucially depends, above all else as discussed above, on two factors: geographic proximity and technological proximity. *Geographic proximity* indicates the positioning of actors within a given spatial framework, while *technological proximity* pertains to the association with the set of vertical or horizontal interdependencies within the scope of production relationships. The transformation of these two types of proximity into a regional system of innovation assumes that they be institutionally organised and structured (Kirat and Lung, 1999). Thus, regional systems of innovation are grounded in collective action at a territorial level. The cohesiveness of such a system of innovation is provided by a spectrum of informal institutions, i.e. the territorially prevailing set of rules, conventions and norms (Kirat and Lung, 1999).

The systems of innovation approach evolved first in a national context (Freeman, 1987), and then in a regional context (see, for example, Cooke et al., 1997; Brazcyk et al., 1998; Malecki and Oinas, 1999). The tradition of studying national systems of innovation has been a recent development (see, for example, Lundvall, 1992; Nelson, 1993, Niosi et al., 1993; OECD, 1994; Edquist, 1997a). Interesting questions and findings have emerged from this literature that sought to establish the extent of convergence and divergence among national innovation systems. This question was of special interest in Europe, given the emergence of European innovation-related institutions that have developed simultaneously with European Community institutions (see Caracostas and Soete, 1997).

It is increasingly being recognised that important elements of the process of innovation become transnational and global, or regional rather than national. The driving forces behind this recognition are two processes that are simultaneously at work today: the process of globalisation of factor and commodity markets and the regionalisation of knowledge creation and learning. This concurs with the view expressed in Ohmae's work on the hollowing-out of the nation state in an increasingly borderless economic world and its identification of the regional rather than the national scale as the relevant economic scale at which leading edge business competitiveness is being organised in practical terms (Ohmae, 1995).

This awareness does not claim that the national scale is unimportant or irrelevant. This scale continues to be crucial in some circumstances. But it is becoming increasingly clear that there is no a priori reason to privilege this particular spatial scale in systems of innovation research, irrespective of time and place (see also Hudson, 1999). A strong case is made today that the regional [i.e. subnational] scale is growing in importance as a mode for innovation systems research. The main argument for this is that regional agglomeration provides the best context for an innovation-based learning economy (Hudson, 1999), for knowledge creation and diffusion and learning. Specific forms of knowledge creation, especially the tacit forms, and of technological learning are both regional and territorially specific. The firms that master knowledge that is not fully codifiable are tied into various kinds of networks with other firms and organisations through regional input-output relations, knowledge spillovers and their untraded interdependencies (Storper, 1997). In some cases market exchange, knowledge spillovers and untraded relations are woven between the various activities within the scope of vertical or horizontal production relationships, but often they are separated.

Formal exchange [i.e. traded interdependencies] and – more importantly – knowledge spillovers and their untraded interdependencies lie at the heart of this line of reasoning:

- *First*, regional input-output relations constitute webs of customer-producer and producer-supplier relations that are essential to communicate information about both technological opportunities and user needs. The user/supplier and producer will gradually develop a common code of communication, making the exchange of information more efficient. To leave a well-established user-producer or producer-supplier relationship becomes increasingly costly and involves a loss of information capital (Lundvall, 1992).
- *Second*, knowledge spillovers occur because knowledge created by one firm or another organisation is typically not contained within that organisation, and thereby creates value for other firms and other firms' customers. Knowledge spillovers are especially likely to result from basic research, but they are also generated from applied research and technological development. This can occur, for example, in obvious ways such as reverse engineering of products, but also in less obvious ones such as when one firm's abandonment of a particular research line signals to others that the line is unproductive and, thus, saves them the expense of learning this themselves. Three vehicles of such spillovers may be distinguished: first, the scientific sector with its general scientific and technological knowledge pool; second, the firm-specific knowledge pool; and, third, the business-business and industry-university relations that make them possible. Once the central role of knowledge spillovers is recognised, a place for informal institutions appears.
- *Third*, untraded interdependencies or regional assets are less tangible benefits that attach to the process of economic coordination and organisational knowledge creation. They are derived from geographical clustering, both economic – such as the development of a pooled labour market – and sociocultural – such as developed routines, shared values, norms, rules and trust that facilitate interactive processes and mutual understanding in the transmission of information and knowledge. Because tacit knowledge is collective in nature and wedded to its sociocultural context, it is more territorially and place specific than is generally thought.

Thus, from a more general perspective it can be argued that it is the combination of territorially embedded Marshallian agglomeration economies, knowledge creation and spillovers and their untraded interdependencies that defines the importance of the regional scale in innovation systems research.

2.6 Some Policy Implications

It is a common view today that the emergence of an innovation-led global economy with all its uncertainties and difficulties of national macroeconomic management is demanding that national governments reconsider their role. In particular, the optimum level of governance is requiring re-examination. Of crucial importance are here the varying competencies of supranational, national and subnational [regional] levels of government within different governance systems such as centralised and federal ones. Matters of macroeconomic management have usually been dealt with at the national level of government. But with increasing significance at supranational governance this is becoming a matter of controversy.

The systemic approach to innovation discussed in the previous sections may provide the conceptual basis for developing particular forms of public intervention, with special emphasis on increasing the regional capabilities for innovation. Edquist and Rees (2000) suggest to distinguish two types of policy implications of the approach. *First,* the systems of innovation approach contains *general policy* implications that can be extracted from the characteristics of the approach. *Second,* the approach provides a framework for identifying *specific policy* issues that should become the focus of policy actions.

General policy implications of the approach are related to organisations and institutions, relations between these, lock-in situations and demand side instruments. A general policy implication derived from the interactive character of the innovation process is that innovation policy should not focus only on the organisations and institutions, but also and – perhaps foremostly – on the links and relations between them. Interactions between the different components of a regional innovation system can be hampered by a number of factors: *first,* conflicting incentive structures [cultures] of partners, for example, university versus business; *second,* market failures that may impede access by firms [especially smaller ones] to information and knowledge, or may distort firm's incentives to invest in technology or absorption capacity [for example, high transaction costs, lack of competition stimuli]; *third,* lack of managerial competencies reducing the understanding of the role of innovation in competitive strategy, and of the benefits of collaborative strategy of knowledge acquisition; *fourth,* financial markets that are unable to assess firm's investment in interactive learning [for example, underdevelopment of specialised venture capital], last but not at least, lack of human resources on which to build capabilities to absorb external knowledge (Edquist and Rees, 2000).

Hence, the interactions between the organisations and institutions should be facilitated by means of policy if they are not functioning sufficiently well. In periods of structural change, regions might have to redesign many of its organisations and institutions as well as the interface between them. The interdependent and non-linear view that characterises the systems of innovation approach implies that it is natural to bring in demand as an important determinant of innovation. This widens traditional supply-side diffusion measures towards policies that recognise innovation and diffusion as interdependent processes. This translates into a greater role for demand driven programmes, network-building initiatives, measures to upgrade the technology diffusion infrastructure and improve its relevance for and accessibility by smaller firms. In order to achieve maximum leverage technology diffusion policies have to build on existing interrelationships in regional innovation systems, especially networks within which firms collaborate and exchange tacit forms of knowledge.

The evolutionary character and the path dependence of innovation processes delivers arguments for early policy interventions and for supporting the emergence of new sectors that would facilitate transitions from dead-end trajectories for regions and firms. Negative 'lock-in' situations imply a role for policy in adapting the shifts in technology and demand (Edquist and Rees, 2000).

In addition, the systems of innovation approach can fruitfully serve as a framework for developing specific policy issues. Concrete comparative analyses are necessary for the design of specific policies in the field of innovation. It is appropriate because it places innovation at its very centre and enables to capture differences between systems of innovation and to identify these problems that should be object of policy intervention. This might indicate to policy makers when, where and how to use financial resources for innovation purposes. It might also indicate how to devise institutions and organisations, how to redesign the interfaces etc. (Edquist and Rees, 2000).

References

Andersson Å. (1985): Creativity and Regional Development, *Papers of the Regional Science Association* 56, 5-20

Anselin L., Varga A. and Acs Z. (1997): Local Geographic Spillovers between University Research and High Technology Innovations, *Journal of Urban Economics* 42, 422-448

Archibugi D. (1992): Patenting as an Indicator of Technological Innovation: A Review, *Science and Public Policy* 19, 357-368

Bienaymé A. (1986): The Dynamics of Innovation, *International Journal of Technology Management* 1, 133-159

Braczyk H.-J., Cooke P. and Heidenreich M. (eds.) (1998): *Regional Innovation Systems. The Role of Governances in a Globalized World*, UCL Press, London

Breschi S. and Malerba F. (1997): Sectoral Innovation Systems: Technological Regimes, Schumpeterian Dynamics, and Spatial Boundaries. In: Edquist C. (ed.) *Systems of Innovation. Technologies, Institutions and Organizations*, Pinter, London, pp. 130-156

Caracostas P. and Soete L. (1997): The Building of Cross-Border Institutions in Europe: Towards a European System of Innovation? In: Edquist C. (ed.) *Systems of Innovation. Technologies, Institutions and Organizations*, Pinter, London, pp. 395-419

Carlsson B. (ed.) (1995): *Technological Systems and Economic Performance: The Case of Factory Automation,* Kluwer, Dordrecht

Castells M. (1996): *The Rise of the Network Society*, Blackwell, Oxford

Chesnais F. (1988): Technical Cooperation Agreements Between Firms, *STI Review* 4, Organisation for Economic Cooperation and Development, Paris

Cohen W.M. and Levinthal D.A. (1989): Innovation and Learning: The Two Faces of R&D, *Economic Journal* 99, 569-596

Cooke P. (1998): Introduction: Origins of the Concept. In: Braczyk H.-J., Cooke P. and Heidenreich M. (eds.) *Regional Innovation Systems. The Role of Governances in a Globalized World*, UCL Press, London, pp. 2-25

Cooke P., Gomez Uranga M. and Etxebarria G. (1997): Regional Innovation Systems: Institutional and Organisational Dimensions, *Research Policy* 26, 475-491

DeBresson C. and Amesse F. (1991): Networks of Innovators: A Review and Introduction to the Issue, *Research Policy* 20, 363-379

Dicken P. (1998): *Global Shift. Transforming the World Economy*, The Guilford Press, New York [3rd edition]

Dunning J.H. (1997): Technology and the Changing Boundaries of Firms and Governments. In: OECD (ed.) *Industrial Competitiveness in the Knowledge-Based Economy: The New Role of Governments*, Organisation for Economic Cooperation and Development, Paris, pp. 53-67

Edquist C. (1997a): *Systems of Innovation. Technologies, Institutions and Organizations*, Pinter, London

Edquist C. (1997b): Systems of Innovation Approaches – Their Emergence and Characteristics. In: Edquist C. (ed.) *Systems of Innovation. Technologies, Institutions and Organizations*, Pinter, London, pp. 1-35

Edquist C. and Johnson B. (1997): Institutions and Organizations in Systems of Innovation. In: Edquist C. (ed.) *Systems of Innovation. Technologies, Institutions and Organizations*, Pinter, London, pp. 41-63

Edquist C. and Rees G. (2000): *Learning Regions and Cities: Learning in Regional Innovation Systems – A Conceptual Framework,* Paper Presented at the International Workshop on '*Knowledge, Complexity and Innovation Systems*', Vienna, July 1-3, 2000

Fischer M.M. (1999): The Innovation Process and Network Activities of Manufacturing Firms. In: Fischer M.M., Suarez-Villa L. and Steiner M. (eds.) *Innovation, Networks and Localities*, Springer, Berlin, Heidelberg, New York, pp. 11-27

Fischer M.M. (2001): Innovation, Knowledge Creation and Systems of Innovation, *The Annals of Regional Science* 35 (2), 199-216

Fischer M.M. (2002): The New Economy and Networking. In: Jones D.C. (ed.) *Handbook of Economics in the Electronic Age*, Academic Press, San Diego [in press]

Fischer M.M. and Varga A. (2001): Geographic Knowledge, Spillovers and University Research: Some Evidence from Austria. In: Institute for Geography and Regional

Research, University of Vienna (ed.) *Geographischer Jahresbericht aus Österreich*, Vienna, pp. 37-47
Fischer M.M., Revilla Diez J. and Snickars F. (2001): *Metropolitan Innovation Systems. Theory and Evidence from Three Metropolitan Regions in Europe*, Springer, Berlin, Heidelberg. New York
Fischer M.M., Suarez-Villa L. and Steiner M. (eds.) (1999): *Innovation, Networks and Localities*, Springer, Berlin, Heidelberg, New York
Freeman C. (1986): The Role of Technological Change in National Economic Development. In: Amin A. and Goddard J. (eds.) *Technological Change, Industrial Restructuring and Regional Development*, Allen & Unwin, London, pp. 100-114
Freeman C. (1987): *Technology and Economic Performance: Lessons from Japan*, Pinter, London
Freeman C. (1991): Networks of Innovators: A Synthesis of Research Issues, *Research Policy* 20, 499-514
Gregersen B. and Johnson B. (1997): Learning Economies, Innovation Systems and European Integration, *Regional Studies* 31(5), 479-490
Guinet J. (1997): Knowledge Flows in National Innovation Systems. In: OECD (ed.) *Industrial Competitiveness in the Knowledge-Based Economy*, Organisation for Economic Cooperation and Development, Paris, pp. 173-178
Håkansson H. (1987): *Industrial Technological Development: A Network Approach*, Croom Helm, London
Hall P. (1986): The Theory and Practice of Innovation Policy. An Overview. In: Hall P. (ed.) *Technology Innovation and Economic Policy*, St. Martin's Press, New York, pp. 1-34
Hudson R. (1999): The Learning Economy, the Learning Firm and the Learning Region: A Sympathetic Critique of the Limits to Learning, *European Urban and Regional Studies* 6(1), 59-72
Jaffe A.B. (1996): Economic Analysis of Research Spillovers. Implications for the Advanced Technology Program, Prepared for the Advanced Technology Program, Brandeis University and National Bureau of Economic Research
Jaffe A., Trajtenberg M. and Henderson R. (1993): Geographic Localisation of Knowledge Spillovers as Evidenced by Patent Citations, *Quarterly Journal of Economics* 63 (3), 577-598
Johannisson B. (1991): University Training for Entrepreneurship: Swedish Approaches, *Entrepreneurship and Regional Development* 3, 67-82
Johnson B. (1997): Systems of Innovation: Overview and Basic Concepts. In: Edquist C. (ed.) *Systems of Innovation. Technologies, Institutions and Organizations*, Pinter, London, pp. 36-40
Karlsson C. and Manduchi A. (2001): Knowledge Spillovers in a Spatial Context – A Critical Review and Assessment. In: Fischer M.M. and Fröhlich J. (eds.) *Knowledge, Complexity and Innovation Systems*, Springer, Berlin, Heidelberg, New York, pp. 101-123
Kirat T. and Lung Y. (1999): Innovations and Proximity. Territories as Loci of Collective Learning Processes, *European Urban and Regional Studies* 6(1), 27-38
Kline S.J. and Rosenberg N. (1986): An Overview of Innovation. In: Landau R. and Rosenberg N. (eds.) *The Positive Sum Strategy*, National Academy Press, Washington, pp. 275-305
Lundvall B.-Å. (1988): Innovation as an Interactive Process: From User-Producer Interaction to the National System of Innovations. In: Dosi G., Freeman C., Nelson R.,

Silverberg G. and Soete L. (eds.) *Technical Change and Economic Theory*, Pinter, London, pp. 349-369
Lundvall B.-Å. (ed.) (1992): *National Systems of Innovation: Towards a Theory of Innovation and Interactive Learning*, Pinter, London
Malecki E.J. (1997): *Technology & Economic Development. The Dynamics of Local, Regional and National Competitiveness*, Addison Wesley, Longman, Essex
Malecki E.J. and Oinas P. (eds.) (1999): *Making Connections. Technological Learning and Regional Economic Change*, Ashgate, Aldershot
Malecki E.J., Oinas P. and Ock Park S. (1999): On Technology and Development. In: Malecki E.J. and Oinas P. (eds.) *Making Connections. Technological Learning and Regional Economic Change*, Ashgate, Aldershot, pp. 261-275
Mansfield E., Romeo A., Schwartz M., Teece D., Wagner S. and Brach P. (1982): *Technology Transfer, Productivity, and Economic Policy*, Norton, New York
Maskell P. (1999): Globalisation and Industrial Competitiveness: The Process and Consequences of Ubiquitifications. In: Malecki E.J. and Oinas P. (eds.) *Making Connections. Technological Learning and Regional Economic Change*, Ashgate, Aldershot, pp. 35-39
Maskell P. and Malmberg A. (1999): The Competitiveness of Firms and Regions. 'Ubiquitification' and the Importance of Localized Learning, *European Urban and Regional Studies* 6 (1), 9-25
Morgan K. (1997): The Learning Region: Institutions, Innovation and Regional Renewal, *Regional Studies* 31 (5), 491-503
Myers M.B. and Rosenbloom R.S. (1996): Rethinking the Role of Industrial Research. In: Rosenbloom R.S. and Spencer W.J. (eds.) *Engines of Innovation: US Industrial Research at the End of an Era,* Harvard Business School Press, Boston, pp. 209-228
Nelson R.R. (ed.) (1993): *National Innovation Systems. A Comparative Analysis*, Oxford University Press, New York, Oxford
Nelson R.R. and Winter S. (1982): *An Evolutionary Theory of Economic Change*, Harvard University Press, Cambridge [MA]
Noisi J., Saviotti P., Bellon B. and Crow M. (1993): National Systems of Innovation. In Search of a Workable Concept, *Technology in Society* 15 (2), 207-227
Nonaka I. and Takeuchi H. (1995): *The Knowledge-Creating Company. How Japanese Companies Create the Dynamics of Innovation*, Oxford University Press, Oxford, New York
OECD (1992): *Technology and Economy: The Key Relationships*, Organisation for Economic Cooperation and Development, Paris
OECD (1994): *National Systems of Innovation: General Conceptual Framework*, DSTI/STP/TIP 94(4). Organisation for Economic Cooperation and Development, Paris
OECD (1995): *National Systems for Financing Innovation*, Organisation for Economic Cooperation and Development, Paris
OECD (1996a): *Employment and Growth in the Knowledge-Based Economy*, Organisation for Economic Cooperation and Development, Paris
OECD (1996b): *Technology, Productivity and Job Creation*, Organisation for Economic Cooperation and Development, Paris
Ohmae K. (1995): *The End of the Nation State*, Free Press, New York
Oinas P. and Malecki E.J. (1999): Spatial Innovation Systems. In: Malecki E.J. and Oinas P. (eds.) *Making Connections, Technological Learning and Regional Economic Change,* Ashgate, Aldershot, pp. 7-33
Pavitt K. (1984): Sectoral Patterns of Technical Change. Towards a Taxonomy, *Research Policy* 13, 343-373

Polanyi M. (1966): *The Tacit Dimension*, Routledge, London
Porter M.E. and Fuller M.B. (1986): Coalitions and Global Strategy. In: Porter M.E. (ed.) *Competition in Global Industries*, Harvard Business School Press, Boston, pp. 315-343
Powell W.W. (1990): Neither Market nor Hierarchy: Network Forms of Organization. In: Staw B.M. and Cummings L.L. (eds.) *Research in Organizational Behavior*, JAI Press, Greenwich, CT, pp. 295-335
Radding A. (1998): *Knowledge Management: Succeeding in the Information-Based Global Economy,* Computer Technology Research Corp., Charleston
Romer P. (1990): Endogenous Technical Progress, *Journal of Political Economy* 98, 71-103
Saviotti P.P. (1988): Information, Entropy and Variety in Technoeconomic Development, *Research Policy* 17, 89-103
Saviotti P.P. (1998): On the Dynamics of Appropriability of Tacit and of Codified Knowledge, *Research Policy* 26, 843-856
Saviotti P.P. (2001): Networks, National Innovation Systems and Self-Organisation. In: Fischer M.M. and Fröhlich J. (eds.) *Knowledge, Complexity and Innovation Systems*, Springer, Berlin, Heidelberg, New York, pp. 21-45
Storper M. (1997): *The Regional World. Territorial Development in a Global World*, The Guilford Press, New York, London
Teece D.J. (1981): The Market of Know-How and the Efficient Transfer of Technology, *Annals of the American Academy of Political and Social Science* 458, 81-96
Teece D.J. (1986): Profiting from Technological Innovation: Implications for Integration, Collaboration, Licensing and Public Policy, *Research Policy* 15, 285-305
Tijssen R.J.W. (1998): Quantitative Assessment of Large Heterogeneous R&D Networks: The Case of Process Engineering in the Netherlands, *Research Policy* 26, 791-809

3 Regional Industrial Policy and the New Economy

Riccardo Cappellin
University of Rome Tor Vergata, Department of Economics and Institutions

3.1 Introduction

The experience of recent decades in the European Union shows that the development of a regional productive system depends in a crucial way not only on the endogenous resources and abilities, but also on relationships with the international economy and the intensity of the integration with other regional production systems in terms of the flow of goods, production investments, financial investments or acquisitions of enterprises, the transfer of organizational abilities, the transmission of technological know-how and tourist flows.

The empirical evidence contradicts the traditional paradigm, according to which a better economic integration of the economically lagging regions with those more developed should determine mechanisms giving rise to a cumulative increase in the interregional disparities. Instead, the changes that have occurred in the technologies and in the forms of the relationships between enterprises seem to show that greater integration at a European level has been a factor that has promoted a more rapid development of the economically lagging regions.

While according to the principle of the economies of scale the regional policies should concentrate resources in areas of very limited dimensions, the positive effects of network economies increase relatively to the increase in the number of regions that are participating in a European Union programme, and this points to the importance of enhancing the integration especially between adjacent areas. In fact, the isolation of a single region with respect to the adjoining areas prevents it from reaching the critical point at which it becomes involved in the context of increasing competition at a global level, and it is necessary for each area to develop a strategy of cooperation at least with neighbouring regions.

This contribution aims, first of all, to illustrate the model of economic development that has characterised the most industrialised regions of the European Union during the last decade. It therefore analyses the conditions that must be guaranteed for favouring the extension of such a model to the peripheral and less developed regions. These considerations have important implications for a strategy of industrial development for the peripheral and economically lagging regions, based on the promotion of the integration of production with the most developed and central regions.

3.2 Models of Industrial Organisation

The model of industrial organisation that emerges on the threshold of the 21st century is profoundly different from that of mass industrialisation, on which traditional economic theory is based. In a modern industrial economy, the model of industrial organisation based on the concept of economies of scale has been replaced by a new organisational model based on increasing integration, cooperation and competition between different firms belonging to the same sector of activity.

Increasingly, the speed at which product innovations are adopted and the flexibility of the integration with other local and foreign firms or organisations are crucial factors in the competitiveness of firms, even more than the continuous expansion of production capacity, through investment, or the containment of production costs by the imposition of lower wage levels. The most dynamic regions are those where the negative impact on the employment levels determined by the inevitable closures and by the downsizing of the existing firms is compensated by the starting up and the fast growth of many small and medium sized firms, often in new sectors of activity.

Although large and small firms represent two alternative models of production organisation, they must face a common competitive challenge and they follow complementary organisational approaches. In fact, the process of the search for better flexibility on the part of the large firms, that adopt a more decentralised structure, corresponds to the process of the search by the small firms for forms of integration through networks, which may have a rather formalised and stable character. Thus, the traditional paradigm of the large firm, which is usually adopted in modern business studies or in the traditional neoclassical theory, differs in several respects from the model of the networks of firms, as shown in Table 3.1.

In fact, those actors, that is to say the active parties involved, who behave according to an entrepreneurial logic have an increasing importance compared to those which behave according to a managerial logic. Similarly, the flexibility of the complex relationships of financial partnership existing in a network of different firms becomes more stable and based on financial control, as is usual within a financial group. Within a network, competitiveness is not so well ensured by the decrease of labour costs and by the flexibility or decrease in the number of employees, as it is by the systematic decrease in the prices paid to the suppliers and by the high flexibility of the contractual relationships with these latter. That flexibility contrasts with the relatively stable relationships, based on reciprocal trust and loyalty, with the workers, or at least with the most skilled workers, within each firm.

Table 3.1 Two models of industrial organisation

Classical Model	**Network Model**
Manager	Entrepreneur
Stable financial control relationships in subsidiaries	High variability of the financial participation both within the group and between the different financial groups
Decrease of labour costs	Decrease of prices paid to subcontractors
Flexibility of labour and layoff of employees	Labour stability and loyalty and flexibility of subcontracting agreements
R&D investments in specific technological fields	Integration of different technologies and continuous investment in training
Incremental expansion of production capacity through investments in new production plants	Fast technological obsolescence and productivity increase through restructuring and downsizing
Maximisation of sales and share in the national market	Downsizing and focus on specific segments of the international market
Exploitation of 'scope economies' between different products	Focus on specific products and acquisition of complementary technologies
Increase in sales and in size of firm	Spin-off of new ventures for the development of new products
Growth of sales through exports and exchange of know how	Internationalisation of the firm
Greenfield investments in new countries and regions	Acquisition of and cooperation with local firms in new countries and regions

Technological progress is not only ensured by incremental investments in R&D within a specific scientific field, but also by the integration of the different technologies of the various firms belonging to the network and by a major investment in the continuous training of the workers. While individual firms seem to aim for an incremental expansion of production and for an increase in productive capacities through investment in new plants, within a network the crucial objective becomes to promote the speed of the substitution of the existing production equipment, which is subject to an increasingly rapid technological obsolescence. This implies the objective of reducing the internal production capacity of the firms through the downsizing and the outsourcing process.

The firms in a network aim at increasing their own size not through the continuous growth of market share in their respective fields of traditional production, but through the spin-off of new production into new firms, through financial links. Finally, while individual firms aim to increase exports and to invest in the creation of new production plants in foreign countries, a network organisation leads to greater importance being assigned to the internationalisation process of the individual firm, the acquisitions of foreign firms and the various forms of cooperation with foreign firms, aimed at promoting the exchange of know-how.

Therefore, it is necessary to carry out a thorough analysis of various mechanisms of crucial importance in the production organisation of modern industry, such as:

- the mechanisms of financial control and the process of strategic decision making,
- the mechanisms of decision making with regard to labour demand and labour organisation,
- the processes of production growth and decisions regarding investments,
- the mechanisms related to the growth of productivity, the decrease of costs and the adoption of process innovations,
- the processes of adoption of product innovations or of product diversification,
- the processes of expansion on an international scale or of internationalisation.

This analysis demonstrates that these mechanisms are very different, when the perspective of analysis is enlarged from the case of the individual firm, which vertically integrates the various phases of a given production process, to the case of a system or a network of firms, which are, however, relatively independent of each other.

Therefore, the increasing integration of the firms in sectoral or geographical clusters makes increasingly inadequate the methodologies of analysis which aim at analysing mechanisms exclusively according to a traditional micro-economic or business approach. Rather, it indicates the need for new methods of analysis capable of considering the mechanisms that operate at the level of an overall system of firms or that have a meso-economic or intersectoral character.

Moreover, it is not certain that the integration of production in a single large firm would involve an organisation, which may be more efficient in static terms and especially more robust, flexible and competitive in the medium and long term with respect to the case of the local production systems or of the sectoral clusters, which are usually composed of about a thousand small and medium sized firms and have an overall work force of around ten thousand employees, in the same sector or in sectors, which are tightly integrated with each other.

In particular, it is clear that the tools of regional or national industrial policy, that have been elaborated in order to operate on individual firms, like the various tools of financial incentives, turn out to be absolutely inadequate in the case of the networks or of the groups composed of a great number of both large and small firms. Certainly, they appear at the most of limited effectiveness, unless they are integrated with other tools which aim at creating a system and which operate on the mechanisms of relationship that link the different firms together in a local production system characterised by strong internal integration. Industrial policy should pass from a distributive logic, using financial incentives, to a logic of regulation/deregulation. This also implies a shift from a prescriptive policy, that in a dirigiste manner points out specific productions or technologies, to a transactive policy (Cappellin, 1998), that acts on the transaction costs or on the relationships of integration between the individual firms, and which facilitates the various technological and organisational changes in a given geographical or sectoral system of firms.

3.3 Innovation and Knowledge Production

The mechanism through which scientific and technological knowledge is produced is radically changing (Lundvall and Johnson, 1994; Mansel and Wehn, 1998; Rubenson and Schuetze, 2000) and this also requires a reorientation of the support policies of the institutions engaged in R&D. According to the linear model, the innovative process of the enterprises is, in oversimplified terms, a serial activity that goes from basic research to applied research, design, production and, eventually, to product commercialisation. Instead, Kline and Rosenberg suggested that innovative processes can be better understood as integrated processes, where different functions, pieces of knowledge, individuals and organisations continuously interact with each other.

In more general terms a new interpretation is emerging which focuses not so much and not only on the mechanisms for the transfer and exchange of know-how, but also, and more specifically, on the processes of know-how integration. According to this interpretation, which is closely related to the development of cognitive sciences, learning processes are described as constructive processes, which imply not so much and not only a mere acquisition of new information, but rather processes concerning the re-formulation of problems, the construction of different categories or mental models of representation, the recombination and development of new heuristic procedures for problem solving.

In this new model the starting point is the formulation of the problems and the search is carried out according to a problem-solving approach with complex interaction between experts, users and financiers. In fact, in many advanced areas of research, numerous competencies are required in order solve the problems. Hence the most important research issues and results turn out not to be those emerging from one individual discipline, but rather those resulting from the new, original integration of traditionally unrelated disciplinary approaches.

The advancement of knowledge is promoted by cultural diversity, and the process of innovation is fostered by the combination of different already existing technologies. Thus, an important distinction in the process of innovation is that between specialised knowledge and diversified knowledge. The first develops within a specific production sector, a given technological trajectory or within a specific world of specialists. The second implies the availability and the integration of the technological and organisational knowledge derived from other sectors and regions. This requires the capability to understand and to translate languages and models, which may be very different from each other.

The production of knowledge is tightly linked to the learning process by the users and the producers. Codified and tacit forms of knowledge are combined in the generation of new knowledge. For the use of knowledge and the introduction of innovation, the process of gradual development of the technological capabilities of the users is necessary. For the producers of knowledge, the interactive process of learning by means of innovation is important. Therefore, the processes of diffusion and of the production of knowledge are tightly connected (Nonaka and Konno, 1998).

The distinction between information and knowledge is based on the fact that information implies the ability to emit, circulate and receive of a flow of messages, while knowledge refers above all to the individual action that gives impetus to a process of comprehension of information received and requires the mechanisms of learning. The processes of learning have an interactive character, and concern both the individual and groups of individuals, both within a firm and outside it [networks]. Often the organisation of a creative process extends beyond the borders of the firm itself, since the associated partners within the enterprise are the suppliers, the customers, and even the competitors, in the case of a cooperative.

The complexity of learning could be correlated to the importance of the interactions that this supposes and upon which it depends. The more complex the process of learning, the more often it requires interaction. Because of the interactive character of learning, the development of the ability to communicate between the agents and to create knowledge is a crucial factor for the dynamic efficiency of the economy.

Certainly, the production of scientific knowledge is no longer exclusive to special institutions, like universities or public research institutes, from which the knowledge should spread like a spillover or a spin-off to the benefit of other sectors. In present day society, based in fact on knowledge, the process of knowledge production has a spatial and institutional dimension. The number of sites and of actors actively involved in generating know-how is growing at an exponential rate. Included among these sites and actors there are also informal groups, networks and associations, consulting firms, and small innovative firms funded by venture capital.

In this new model, the most advanced types of research are carried out in the context of specific applications and on the basis of intense interaction between the producer and the user of knowledge. Therefore, there is a tendency for this new knowledge to have an increasingly 'private' character, since it is not elaborated by institutions obliged to reveal it free of charge, such as public RTD institutions, but by organisations, such as industrial firms, that do not divulge the knowledge but keep it for their own use with the objective of putting it to profit,.

The processes of innovation are defined as processes for the creation of new knowledge or the combination of knowledge which already exists. Therefore, the organisation of the production of knowledge is affected by changes that could have deep consequences for the traditional established institutions and for managing organisational change. In particular, the way in which scientific and technological know-how is produced demands a profound change in the policies aimed at supporting R&D institutions.

In this framework incubators represent a particularly interesting experience. The notion itself of incubators has traditionally emphasised the physical aspect, usually representing them as buildings or as specific sites destined to host new enterprises for a pre-established period of time, with the main objective of either revitalising redundant industrial buildings or creating an industrial district accommodating small, not necessarily innovative firms. Examples of these kinds of incubators are frequently found in Italy and in many other European countries. On the contrary, far from being important for their physical characteristics, modern incubators aim at enhancing the technological and organisational know-how of new firms. In particular, a modern incubator is not only characterised by the being provided with adequate infrastructures, services and sites, but above all by its access to a diversified network of qualified services, which is of crucial importance. A modern incubator should also help the new entrepreneur in the activity of fund raising, not only of public finances, but also of private resources, such as those required at the beginning and before the creation of the firm [seed capital] and the venture capital resource, required in the first phases of the development of the firm. It also helps the new firm in its search for industrial groups which may consider financial participation in the new firm as a useful instrument for the development of new technologies and of new products and services.

What is more important, is that the new incubators are no longer meant as mere providers of services to other enterprises, but may rather represent fully fledged enterprises, which grows and develops as their know-how [derived from the promotion of different business plans] grows. This new kind of incubator has become more common in the United States and in Israel in the last five to ten years (cf. www.cambridgeincubator.com/about/faq.htm; www.incubators. org.il/).

3.4 The Impact of the New Economy on the Organisation of Industry

The new economy is composed of at least two different sectors. The first is represented by the production of investment goods that incorporate information and communications technology [ICT]. This sector is concentrated mainly in the Anglo-Saxon countries and in those of the Far East. It is a sector that will be characterised by a big turnover of firms and progressively lower barriers to entry, as soon as the technology spreads. The second component of the new economy is represented by the use of the ICT networks and particularly of internet in the different sectors of the economy, both industrial and tertiary. In this framework an extremely important role is played by the network economies that strengthen the advantage of the firms as the number of users increases. This determines barriers to entry and the development of cumulative processes that benefit those firms that are the first innovators. This second sector represents a big opportunity for many European countries, above all from the perspective of strengthening the existing competitive advantages in the different industrial sectors of specialisation, such as mechanical, chemical and textile production.

According to a largely diffused interpretation, internet would allow a better circulation of information regarding tenders or the creation of electronically specialised markets, and it would lead to more intense competition between firms and to an increasing downward pressure on the profit margins of the producers of components and of commodities. This should imply a decrease of costs that may vary from sector to sector, but that could range from five to 15 percent in telecommunications up to as much as 30 percent in the case of electronic components.

Considering the situation from a different point of view, it seems that if the effect of internet will be that of reducing transaction costs in sales and purchases between firms, this should favour the development of cooperation between customer and supplier and stimulate a process of outsourcing, rather than promote forms of vertical integration. The e-economy seems to lead more to spin-offs than to mergers of firms. Moreover, according to this alternative model, the greater effects of internet are not so much connected to the circulation of the flows of goods in relationships of supply, but to the reduction of the costs of transaction in the circulation of information. Therefore, the use of internet accelerates the circulation of information, it favours cooperation between different actors even over long distances, and it makes the decision-making processes faster. The use of internet makes time competition based on innovation more important than price competition, and it works on the so-called adjustment costs, that hinder or facilitate technological, organisational and institutional change. This is perhaps the more remarkable effect of internet and certainly the more pervasive effect at the intersectoral level.

The advantages of internet should not be sought so much in the reduction of costs, as in an increase in the effectiveness of the relationships between firms and in an improvement in the quality of the product-services of the individual suppliers, and in a better economic performance of the buyers. In fact, the ability to react quickly, to make use of adequate resources at the appropriate moment and to find the most competent partner as quickly as possible is crucial. In such a context, to know how to communicate, cooperate and interact becomes much more important than it once was. (Maillat and Kebir, 1999).

Particularly, the acceleration of technological progress and the tendency for the flow of information to move at an ever increasing speed leads firms to adopt organisational forms which are characterised by greater stability or by relationships of semi-integration, that allow the costs of coordination and of transaction to be reduced, with respect to market or commercial relationships based on competition and conflict.

The development of the possibilities for interaction between actors requires an increasing formalisation of the relationships between firms, the adoption of specific codes, and the creation of computer networks, as well as revision of the internal organisational models. The use of internet will involve investment in organisational procedures and know-how and in the training of workers, that will bind together the partners in an increasingly tight relationship.

There exists a direct relationship between the speed of the flows and the stability of the nodes in a network. The shift from market or commercial relationships to relationships integrated within a network, on one hand requires investment in new hard and soft infrastructures by individual firms, and on the other hand allows a faster circulation of information between these same firms. In fact, the production systems of small and medium-sized firms of many South European regions are characterised by the explicit and systematic search for greater stability, both in labour relationships and in the relationships with customers and with suppliers. This assures greater flexibility in face of the technological changes and decreases marketing times.

Therefore, internet represents a crucial opportunity for a more effective internal integration of the network systems of small and medium sized firms that characterise many South European regions. Internet allows a huge increase in the speed of the acquisition of orders and in the reduction of stocks. It promotes a general reorganisation of the production cycles both in supplier and in buyer firms, and it allows a drastic reduction of errors and an improvement in the quality of processing. Internet, moreover, allows the development time of new products to be reduced and facilitates a true relationship of co-production between the buyer and the supplier.

3.5 Processes of Learning and Labour Training

A close integration between firms which exploits the opportunities created by the new ICTs requires a massive renewal of the process of re-engineering. The increased investment of the firms that will use internet will not be therefore so much in technology, but in the processes of organisational learning or in the human resources connected with the process of restructuring. The principal limitation in the diffusion of the new economy will be the lack of qualified workers. It is necessary to ensure better synchronisation between the industrial policies and the policies of the labour market and of labour training.

Since the capability of firms to have access to advanced know-how developed from its buyers or suppliers becomes crucial, it is important for the firms to stabilise the relationships with these latter in order to allow a gradual accumulation and evolution of this know-how. In this picture, the customer-supplier relationship in subcontracting networks appears as a crucial factor in the diffusion of the technological knowledge and of modern methods of production organisation [even if only in time and quality certification]. This leads to a more articulate and organic initiative of training or tutorship, both organised by the buyer for the respective suppliers, and also by the producers of machinery and advanced components for their customers.

The so-called labour flexibility appears in objective contrast to the demand of the enterprises to reduce turnover and to the increasing competition with other firms on the labour market. In fact, the greater complexity, specificity and speed of change of the technologies used in each individual firm and also the complexity of the relationships between the enterprise and an elevated number of suppliers and customers make it increasingly onerous for the firms to replace the workers that would be subtracted from them by the competing firms, capable of offering higher salaries and better working conditions. On the other hand, the pervasive character of the use of the new technology in the different corporate functions seems to underline the necessity for training efforts to be reoriented towards adult workers already employed, and not limited only to the education of the young people entering the labour market.

The retraining of adult workers who are dismissed from the firms involves an enormous effort and has limited effects on the goal of their new integration in the world of work. The problem, instead, is that of promoting the mobility of the workers between different functions during their working life and of favouring a gradual accumulation of competencies through continuous adult training. This implies gradually developing new competencies during the production processes. It requires the development of an attitude in favour of active learning and of the culture of learning. People should in fact consider long term formation processes, recurrent cycles of retraining and adjournment and even continuous education throughout the whole of their working lives as necessary and even normal aspects of economic life. The continuous education of employed workers should therefore become a permanent commitment of firms. This assures their competitiveness, and would contribute towards avoiding the inevitable crises in those firms that for different motives have not renewed their knowledge base. In fact, on one hand the firms do not develop continuous education because they believe they will lose the human resources in which they have invested. On the other hand, the lack of investment in internal human resources leads the firms to subtract qualified personnel from the other firms in the same sector. More and more this represents one of the motives for which the more motivated workers voluntarily leave their firms and for which the young workers are not attracted by the possibilities of occupation in these firms.

Otherwise, it is inevitable that the labour market will be increasingly characterised by a shortage of workers with those qualifications for which the investment both of the firms and of the training institutions has been inadequate, and also by an increase in the long-term unemployment of adult workers.

3.6 The Territorial Network Approach

During the last decade, regional science literature has focused on the process of the adoption of innovations in local production systems and has elaborated various concepts, such as those of the milieu innovateur, the regional innovation system, the learning regions, the dynamics of proximity or the institutional consistency. An advantage of these concepts is that they can be applied to a broad scope of types of local production systems, such as the traditional industrial districts (Becattini, 1991; Garofoli, 1991; Brusco and Paba, 1997) and also to the less developed regions, where specialised industrial districts have not yet appeared, or the most developed regions, which seem to have overcome the phase of a tight sectoral specialisation. In particular, an analysis of this recent literature allows the identification of various concepts, which seem to be rather useful in order to explore the regional or territorial dimension of the new economy or the knowledge economy.

The systemic dimension of innovation: The dynamics of innovation depend on the tight interdependence within the regional system of innovation between a plurality of actors, such as large and small firms, research institutions, higher education institutions, private R&D laboratories, technological transfer agencies, chambers of commerce, entrepreneurial associations, professional training organisations, specific public agencies and appropriate offices of public administration (Cooke, 1998).

Interaction logics and associative governance: For the various local actors to belong to the same regional innovation system represents the base of an associative approach, which leads to explicit cooperation between the various local private, collective and public actors, to the creation of clubs, forums, consortia and different institutional kinds of partnership or of innovation networks (Maillat, 1995; Cooke and Morgan, 1998).

The integration between tacit or implicit knowledge and external codified knowledge: The creation of new knowledge implies an intense interaction process between the tacit or implicit knowledge, linked to the local context, with the codified knowledge, available at a world-wide level, in order to stimulate the regional endogenous potential (Nonaka and Konno, 1998). This requires both face–to- face contacts and physical proximity as well as ICT contacts for long distance communications.

The territorial proximity, considered as geographical proximity, organisational proximity and institutional proximity: A regional innovation system combines these three different notions of proximity (Rallet and Torre, 1998; Bellet et al., 1993). While the organisational proximity deals with the links in terms of the organisation of production, geographical proximity deals with the links in terms of distance.

Institutional proximity consists in the development of relationships having an intentional nature, such as the relationships of cooperation, or of trust, for the exchange of technological information, and of partnerships, which determine the strategies of the actors.

The dynamics of collective learning: In a learning region or in a regional innovation system the firms and the other organisations develop the capability of gradually modifying their behaviour according to change in the external environment and are systematically engaged in interactive learning (Morgan, 1997; Lundvall and Johnson, 1994).

Intermediate institutions and institutional consistency: Institutions are not only constrained by the public sphere, while they emerge from the overall interaction between the individual actors (Rullani, 1998). Moreover, institutional consistency has a precise evolutive character, since the institutional set-up is the result of a long and gradual learning process, and the dynamics of the organisational forms of the system of private firms corresponds to the constant evolution and creation of various organisations and institutions, which integrate and orientate a local production system.

These new concepts lead to the definition of a model of territorial networks, based on the concept of integration and characterised by the capability of establishing a direct link between the industrial/economic dimension and the physical organisation of the territory and the institutional framework of local government. In an industrial perspective, the model of territorial networks implies a greater formalisation of those relationships between the firms, which were mainly informal and based on reciprocal personal knowledge and trust in the traditional industrial districts (Cappellin, 1998). This implies the adoption of more binding agreements between the local firms, as implied by the modern process of 'just in time' and of 'total quality' production organisation.

In fact, a modern local production system is characterised by the specialisation and the complementarity of the firms, and it produces forms of vertical integration, like the filieres, that allow the local production system to combine the continuous evolution of their know-how, that ensures a time sustained competitive advantage, with increasingly more immediate access to the final market, where the added value is created. Thus, a second characteristic of the approach of the territorial networks is a greater sectoral diversification of the local economy and a tight integration among the various sectors.

A third characteristic of the model of territorial networks is the fact that it allows an extension of the geographical framework, in which firms develop their subcontracting relationships, at the interregional and international levels. In fact, the forms of the relationships between the firms and particularly the relationships of subcontracting and of financial integration continually evolve and they extend both to the intersectoral level and to a wider geographical scale (Cappellin and Cersosimo, 1998). A network organisation on an interregional scale allows easier access to information regarding the different outlet markets and the evolution of the needs of the consumers in distant regions and countries. This information may be of an explicit kind and easily available long distance through modern ICT or of an implicit kind available only through the direct or indirect presence in the specific locality considered.

However, the characteristic of territorial embeddedness does not appear to be in contradiction with an increasing external openness at interregional and international levels. On the contrary, there exists close integration between endogenous capabilities and external openness. The concept of a territorial network differs from the traditional concept of the industrial district or of production clusters, since it indicates a shift from an approach of relative selective closure, based on the territorial homogeneity, to an approach based on the concept of territorial integration, which implies the creation of tightly territorially embedded networks, which may have a variable geometry and include both various local actors and various external regions and foreign countries (Cappellin, 1993).

In fact, an organisational model, like that of the territorial network, is capable of promoting the development and continuous change of the knowledge available in the individual local production systems, through the synergy between the internal technologically specialised resources of the local firms and the external technological resources in other regions and countries. Thus, according to this approach a local production system seems to be characterised by the following different types of integration linkage:

- *Technological integration, as indicated* by the development of the local production know-how, the sharing of knowledge and values promoted by learning processes on the job, the continuous education of workers, the vocational education of young workers, joint investment in R&D by local firms, and technological cooperation with external firms,
- *Cooperation between workers and firms and integration of the local labour market,* related to the mobility of workers between firms of the same sector

and also the capability of attracting qualified workers from other regions and from other sectors,

- *Production integration between firms,* through subcontracting relationships between firms which play a crucial role in promoting the diversification of local production,
- *Integration between service sectors and manufacturing firms*, related to the development of modern commercial distribution services, transport and logistic services, and also qualified services regarding the certification of the quality of the production and the diffusion of modern technologies,
- *Financial integration of the firms,* as indicated by the creation of groups composed of several firms belonging to the same entrepreneurial sector and by pro-active bank-industry relationships, which promote the creation of spin-off and the capability of attracting external investment or the investment of local firms in other countries and regions,
- *Territorial integration at a local level,* which requires an improvement in the infrastructures available and promotes effective physical planning aimed at defending the quality of the territory,
- *Social and cultural integration*, which determines the existence of a local identity and brings about consensus throughout a wide spectrum of the local community on a shared strategy,
- *Relationships of institutional integration,* which are related to the development of local administrative capabilities and the capability of local institutions to interact with regional and national institutions in the implementation of strategic development projects,
- *Territorial integration at the interregional and international levels*, which leads to greater openness in an interregional perspective, to the development of a local foreign policy or of a territorial marketing policy, which are crucial in attracting external investment and in promoting the internationalisation of local firms.

Therefore, the model of territorial networks underlines the tight integration between the organisation of economic relationships between firms and the physical organisation of the regional territory, which is characterised by an intense network of industrial settlements, such as industrial districts, and of small and medium size urban centres and by the complex relationships of these latter with the largest metropolitan areas where the modern producer services are concentrated (Cappellin, 1988, 1997, 2000).

Moreover, the model of territorial networks highlights the institutional dimension of the process of local economic development. In fact, in a society based on knowledge or in a learning economy the intervention of the public institutions appears necessary where informative asymmetries exist. The increasing decentralisation of the economy and the increasing complexity of a modern industrial economy create the demand for integration that must be exercised by the public institutions and by new collective organisations. Therefore, the role of local and regional government becomes that of catalysing, like a systems integrator, new solutions which promote the integration of the complementary resources of the different actors, and stimulate the project design ability of local actors, by proposing strategic programmes and also by offering technical assistance in order to carry out specific programmes.

3.7 Policies for Territorial Networks

Local and national policies promoting the development of a modern local production system should aim at three closely inter-related factors of competitiveness, namely (Cappellin and Orsenigo, 2000):

- the speed of change and of the adoption of innovations,
- the forms of networking between the local actors,
- the interregional and international openness of the local economy (see Fig. 3.1).

First, the challenge represented by the speed of change stimulates the local actors to cooperate. Conversely, the cooperation between the local actors facilitates the processes of change. *Second*, the speed of change in the local and regional economy determines the capability of the local production system to sustain the increasing international competition induced by a greater interregional and international openness of the local economy. On the other hand, the openness of the local economy and the possibility of promoting forms of cooperation on an interregional and international scale facilitate the speed of change. *Finally,* cooperation on a local scale represents an important prerequisite for common actions on an interregional scale, while the challenge represented by international competition leads the local actors to undertake forms of cooperation.

A modern regional industrial policy should not adopt sectoral or functional approaches, but it should rather be characterised by a territorial approach. This approach requires a tight integration of the various instruments of industrial policy, having a sectoral or functional character, such as research, internationalisation, training, industrial sites, transport, etc., which are concerned with the same territorial area or the same local production system.

The increasing openness of the local economies and the new forms of economic integration of the firms and the other local actors justify the transition from an approach based on the concept of the economies of concentration and of sectoral specialisation, such as in the traditional industrial districts, to one based on the concept of territorial and intersectoral integration, such as indicated by the network approach. Therefore, regional intervention could involve two different but complementary types of intervention. The first consists of interventions concentrated in individual local production systems. However, it is not necessary to follow the usual procedure that foresees the definition of areas to which the available financial resources should be assigned and therefore the elaboration of development plans and finally of individual projects. Instead, the intervention of the regional government should define in a relatively detailed way specific guidelines or sectoral programmes, that define the criteria of eligibility of the actors and the selection of the individual projects. Therefore, according to a bottom up logic it should involve an ex-post selection of the specific plans, which have been in fact promoted and co-financed by local actors.

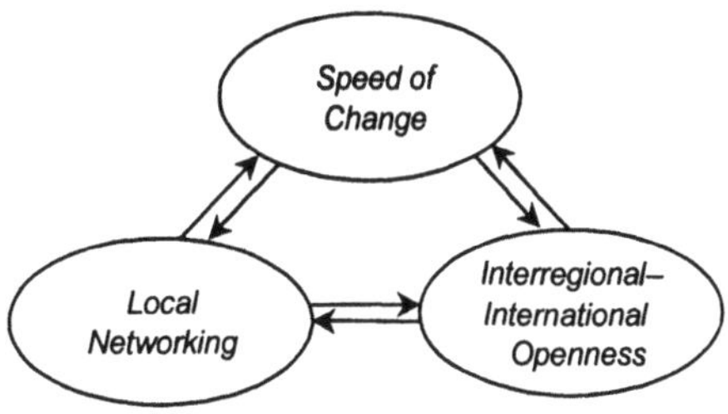

Fig. 3.1 Three factors in the development of a regional production system

The guidelines for the regional administration could, for instance, focus the projects on the following strategic areas, in which strong local orientation seems to be needed and which allow the role of territorial institutions [towns and provinces] to be enhanced:

- strengthening of the processes involved in the formation of new firms and of new entrepreneurship,
- development of the process of diversification and of production reconversion in strongly specialised industrial areas,
- development of production know-how and of continuous professional training within the framework of the European structural policies,
- improvement of territorial quality and the promotion of sustainable development through adequate planning of the territory and of the transport networks.

However, regional industrial policy should not only support the initiatives that aim at strengthening the individual local production systems, but it should also promote actions that have a wider dimension at the regional level, such as for instance the action programmes with the following objectives:

- the integration of the different local production systems at the regional level, starting with those which are adjacent to each other and have a common production role, favouring intersectoral integration and the process of diversification of production,
- investment in R&D in large technologically innovative projects at the regional level,

- the opening up of interregional and international production relationships with production systems located in other regions.

Therefore, this second type of intervention, unlike the first, could be based on a top down logic and promote action programmes with a strong regional emphasis, with a strategic character, and which favour the creation of a regional system of innovation or promote a qualitative jump of the regional production system towards the model of a learning region. In this case the participating actors would not implement isolated projects, but they would participate in an original network focused on a single project. Therefore, the selection of partners from among the local actors should be made ex ante and not ex post, and should be based on a variety of criteria, like the strong embeddedness of such actors in the different sectors of the regional territory and their ability to ensure an original and qualified contribution to the execution of the general project.

3.8 Policy Tools for Interregional Production Cooperation

Policies traditionally consist of financial and fiscal incentives granted to individual enterprises in the less developed regions with the objective of reducing regional disparities, as indicated in Fig. 3.2. However, the fact that the underdeveloped conditions are due to a situation of geographical isolation or of peripherality leads to the promotion of the interregional connections or forms of integration at the interregional level through the creation of transport and communication infrastructures or the promotion of subcontracting agreements at the interregional level (Cappellin, 1999). In both these two cases regional policy aims to act on the competitive advantages in terms of production and transport costs of the firms in the less developed regions.

The consideration of competitiveness in dynamic terms or in terms of the ability to adopt product and process innovations has led to the use of different policy tools. These are the policies involving the creation of service centres for the small and medium-sized firms or of 'real services', of incubators for new innovative enterprises or of technological parks and of relationships between the firms and the research centres. Also in this case, it is opportune to promote a stronger integration between the less developed regions and the more developed regions. Therefore, a policy that aims to promote the cohesion in territorial terms or to integrate the different European regions must aim to promote the construction of a network of firms or strategic alliances and joint ventures at the interregional level.

Interregional production cooperation represents an original approach to the development of a strategy of industrial policy in the economically lagging regions. Contrary to the traditional financial schemes that aim to attract external investments in the less developed regions, initiatives of production cooperation aim to create partnerships between the entrepreneurs of different regions and to spread new organisational models capable of supporting international competition. These forms of interregional cooperation should not follow the traditional objective of establishing individual production plants dependent on the parent enterprises, but should promote the development of a modern 'just-in-time' system of co-manufacturing, based on the outsourcing of components.

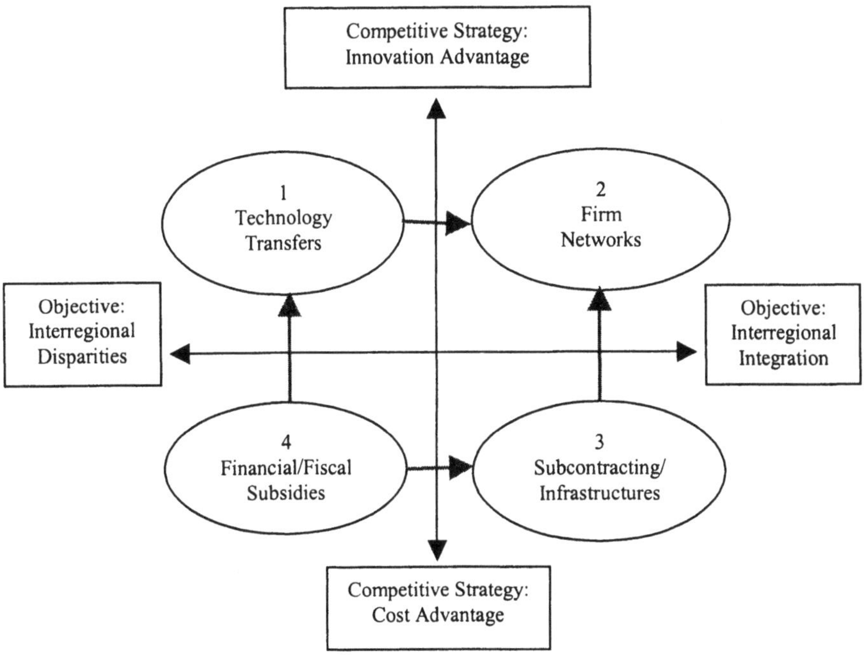

Fig. 3.2 Tools of regional industrial policy

The problem is not only to attract new investment in the economically lagging regions, but instead to also focus on industrial partnership or collaboration between entrepreneurs. This requires the entrepreneurs to be mobile physically as well as financially . In fact, also in the most developed regions of Italy, for example, the dense fabric of artisan enterprises and micro enterprises, which contributes to an explanation of the high rates of industrialisation and the low rates of unemployment, is based on progressively more complex and flexible subcontracting relationships between small firms and between these latter and the medium and large-sized firms, that provide leadership in the respective local production systems.

Tight forms of production integration are now possible also at long distance. In fact we may observe the process of production decentralisation of the small and medium-sized firms of the Central North Italian regions toward the countries from Central East Europe, that in a few years will enter the European Union and are already now moving towards an increasing specialisation in their production and towards integration in the networks of subcontracting carried out by the European firms. Countries like Slovenia, Slovakia, the Czech Republic and even Romania and Albania show remarkable development in this direction. These new schemes of production cooperation should include (Cappellin and Cersosimo, 1998):

- a careful and explicit organisation of the relationships between the groups of firms belonging to the two regions through subcontracting contracts, the offer of logistic services and financial links,
- the interregional mobility of entrepreneurs and of more qualified workers from the more developed to the less developed regions,
- the interregional mobility of the workers of the less developed regions towards the greater establishments of the firms in the most developed regions for purposes of professional training,
- the direct participation and institutional support of the entrepreneurial associations and of the local institutions in the two regions through the creation of appropriate agreements.

The subcontracting agreements must be understood to be a peculiarity of the economically lagging regions in the production of intermediate products [products not aimed at the final consumer] and modern technologies aiming to satisfy the specific needs of industrial customers located in the more industrialised European regions will be adopted. Subcontracting must be framed in a narrow relationship of vertical integration downstream with the external customer firms and upstream with other local subcontracting firms. In fact, production cooperation must also aim to develop the culture of subcontracting in the less developed regions, where firms are usually very tightly vertically integrated. This requires the promotion of a constant search for local subcontractors and the entrepreneurial abilities to be tapped in these regions. The objective is that of creating a network of firms specialised in high quality subcontracting.

3.9 Conclusions

In recent years the regional industrial policies have tended to be based on two models that appear increasingly inadequate. The first model is that of promoting regional champions. The policies aim to strengthen specific highly concentrated productive sectors that have been identified as points of industrial strength in the regional economy. The second approach is that of local development, which has led to the creation of innumerable territorial agreements, local development agencies and local service centres. This approach aims to promote the small entrepreneurs, the development of SMEs and a better integration between firms, territories and intermediate institutions in the different local production systems of the various regions. However, this approach often lacks a strategic vision and is based merely on forms of social animation or on the mere enunciation of institutional partnerships.

On other hand, the recent approach of territorial networks, that has spawned a vast amount of international literature, indicates the need to adopt a systemic approach at the regional level, aimed first of all at the creation of territorial filiere that, especially in the highly specialised territorial areas, promote intersectoral integration at the local level, the diversification and reconversion of production. From this approach it is evident that the general development of the region is not simply the sum of the development of individual local production systems, but it is also the result of their integration and synergy at a regional level.

The development of the new economy and of the modern information and communication technologies on one hand promotes a remarkable increase in the speed of the transformations, and on the other hand requires a greater investment in the nodes of the networks or in the creation of consolidated or stable relationships between the different actors and between the firms and their workers.

The approach of the territorial networks, in fact, underlines the need to integrate the promotion of endogenous development with interregional openness and integration. This shows that the development of modern ICT is not sufficient, whether or not it is accompanied by tools for reducing the organisational and institutional distances, to promote different forms of active and passive internationalisation of the firms and of the individual local production systems. In particular, the development and the internationalisation of the less developed regions require the creation of institutional schemes of interregional cooperation that represent an innovative tool of regional policy.

References

Becattini G. (1991): The Marshallian industrial district as a socio-economic notion. In Pycke F., Becattini G. and Sengenberger E. (eds.), *Industrial Districts and Inter-firm Co-operation in Italy,* International Institute for Labor Studies, Geneva, pp. 37-51

Bellet M. Colletis G. and Lung Y. (1993): Economie de Proximites [Numero Spécial], *Revue d'Economie Régionale et Urbaine* 3, pp. 357-602

Brusco S. and Paba S. (1997): Per Una Storia dei Distretti Produttivi Italiani dal Secondo Dopoguerra agli Anni Novanta, in: Barca F. (ed.), *Storia del Capitalismo Italiano dal Dopoguerra a Oggi*. Roma: Donzelli Editore, pp. 263-333

Cappellin R. (1988): Transaction Costs and Urban Agglomeration, *Revue d'Economie Regionale et Urbaine* 2, pp. 261-278

Cappellin R. (1993): Interregional Cooperation in Europe: An Introduction. In: Cappellin R. and Batey P. (eds.), *Regional Networks, Border Regions and European Integration.* Pion, London, pp. 1-20

Cappellin R. (1997): The Economy of Small and Medium-size Towns in Non Metropolitan Regions, Document Prepared for OECD, *Programme of Dialogue and Cooperation with China*, SG/CHINA/RUR/UA(97)3

Cappellin R. (1998): The Transformation of Local Production Systems: International Networking and Territorial Competitiveness. In: Steiner M. (ed.), *Clusters and Regional Specialization*, Pion, London, pp. 57-80

Cappellin R. (1999): The Enlargement of the European Union and the Future of the European and National Regional Policies, *Regional Contact* 13, pp.29-40

Cappellin R. (2000): Urban Agglomeration and Regional Development Policies in an Enlarged Europe. In Bröcker J. and Herrmann H. (eds.), *Spatial Change and Interregional Flows in the Integrating Europe*, Physica-Verlag, Heidelberg, pp.117-130

Cappellin R. and Cersosimo D. (1998): Sud Chiama Nord: Le Reti Interregionali di Piccole e Medie Imprese. In Donzelli C. (ed.) *Le vie del Mezzogiorno. Storia e scenari*, Meridiana Libri, Lamezia Terme, pp. 213-228

Cappellin R. and Orsenigo L. (2000): The Territorial Dimension of Modern Industry and the Scope of Regional Industrial Policies. In: Klemmer P. and Wink R. (eds.), *Preventing Unemployment in Europe*, Edward Edgard, pp. 166-187

Cooke P. (1998): Introduction: Origins of the Concept, in: Braczyk H. J., Cooke P. and Heidenreich M. (eds.), *Regional Innovation Systems. The Role of Governances in a Globalized World,* UCL Press, London, pp. 2-25

Cooke P. and Morgan K. (1998): *The Associational Economy. Firms, Regions and Innovation,* Oxford University Press, Oxford

Garofoli G.(1991): *Modelli Locali di Sviluppo*, Franco Angeli, Milano

Lundvall B.-Å. and Johnson B. (1994): The Learning Economy, *Journal of Industrial Studies* 1 (2), 23-42

Maillat D. (1995): Territorial Dynamic, Innovative Milieus and Regional Policy, *Entrepreneurship and Regional Development* 7, pp. 157-165

Maillat D. and Kebir L. (1999): 'Learning Region' et Systemes Territoriaux de Production, *Revue d'Economie Regionale et Urbaine* 3, pp. 430-448

Mansell R. and Wehn U. (eds.) (1998), *Knowledge Societies: information technology for sustainable development*, Oxford University Press, Oxford.

Morgan K. (1997): The Learning Region: Institutions, Innovation and Regional Renewal, *Regional Studies* 31 (5), pp. 491-503

Nonaka I. and Konno N. (1998): The Concept of 'Ba': Building a Foundation for Knowledge Creation, *California Management Review* 40 (3), 40-54

Rallet A. and Torre A. (1998): On Geography and Technology: Proximity Relations in Localised Innovation Networks. In: Steiner M. (ed.) *Clusters and Regional Specialization*, Pion, London, pp. 41-56

Rubenson K. and Schuetze H.G. (eds.) (2000): *Transition to the Knowledge Society: Policies and Strategies for Individual Participation and Learning,* Institute for European Studies, Vancouver

Rullani E. (1998): Riforma delle Istituzioni e Sviluppo Locale, *Sviluppo Locale* V (8), 5-46

4 Human Resources and Regional Development

Peter Batey
University of Liverpool, Faculty of Social and Environmental Studies

4.1 Introduction

The systematic development of human resources has for some time been an important consideration for national policy-makers throughout much of the developed world. It is reflected in the increased emphasis that is nowadays given to issues such as lifelong learning, equal opportunities, social inclusion, employability and what has come to be known as the knowledge economy. A recent White Paper published by the UK government is typical of the approach to human resource development that is now taken by policy-makers at a national level:

'*Successful modern economies are built on the abilities of their people. People are at the heart of the knowledge-driven economy. Their knowledge and skills are critical to the success of British business. People are the ultimate source of new ideas. In a fast moving world economy, skills must be continually upgraded or our competitiveness will decline*'. (UK Government White Paper on Competitiveness, see Government, 1998, p. 28).

The same document goes on to stress the importance of developing relevant skills among the workforce and ensuring that as many people as possible are able to compete for jobs:

'[*The*] *lack of skills inhibits innovation and hinders investment. The problem is most acute among smaller businesses which find it difficult to attract good graduates or afford training. But in businesses of all sizes, managers often underestimate the value of skills, including management skill ... In a knowledge-driven economy we cannot squander thirty percent of the potential workforce through poor education and social exclusion*' (Ibid, p. 29).

'*In the global economy, capital is mobile, technology spreads quickly and goods can be made in low cost countries and shipped to developed markets.*

British business therefore has to compete by exploiting capabilities which competitors find hard to imitate. The UK's distinctive capabilities are not raw materials, land or cheap labour. They must be our knowledge, skills and creativity' (Ibid, p. 6).

This concern to maintain and enhance competitiveness has, since the mid-1990s, begun to find expression at a regional level (Batey and Friedrich, 2000). Many regions in developed countries have produced regional competitiveness strategies in which the development of human resources figures prominently. Systems of indicators have been created to measure progress towards targets for the attainment of particular qualifications and skills, and there has been a concern to identify the appropriate roles that can be taken in developing human resources by universities and other education and training institutions.

This chapter presents a survey of current thinking about the human resources component of regional development strategies. It begins by drawing upon a sample of international evidence of variations in education and training levels across the range of OECD countries. The degree of variability on measures such as literacy and participation in tertiary-level education is quite striking, particularly when differences between age groups are taken into account. Though regional variations within a country are unlikely to be as great, this international comparison provides some useful food for thought and a backcloth to the regional comparisons that follow in later sections of this chapter. In Section 4.3 a number of key concepts are reviewed, especially the notions of a learning economy and a learning society and the subsequent translation of these ideas into the concept of a learning region.

Universities are widely seen as having an important contribution to make to regional development and in Section 4.4, the scope for their potential impacts is reviewed. Too often in the past studies of these impacts have taken a rather narrow perspective by focussing purely on the expenditure effects and have ignored the so-called 'knowledge effects' that may ultimately be much more important.

In Section 4.5, examples are considered of how indicators and targets can be employed to chart the progress of human resource development in a system of regions. This leads on, in Section 4.6, to an examination of how human resource development can be introduced as a theme within regional development strategies. Two examples are considered. The first looks at a human resource development strategy for North West England, while the second examines how European and national human resource priorities can be translated into a regional plan for the deployment of European Structural Funds. This latter case study is based on Merseyside, in North West England, a heavily-industrialised region with European Union Objective One status that has historically lagged behind other parts of Europe in terms of education and skill levels and, partly as a result of this, suffers from a high incidence of social exclusion. In the final Section, conclusions are drawn.

4.2 Human Resource Development: Some International Comparisons

In this section a number of international comparisons will be made in order to establish the extent of variations in particular aspects of education and training and their impact upon the workforce. Five topics have been selected to give a broad basis for comparison and in all cases the annual publication of the OECD, *Education at a Glance*, is the source of data. The comparison covers educational attainment, at upper secondary and university [tertiary] levels; literacy; trends in participation in university education; the degree of take-up of continuing education and training; and the impact of university education upon earnings levels.

In comparing the relative levels of education attainment it is instructive to examine both the completion of upper secondary level of education[1] and the completion of an undergraduate level degree, within tertiary education[2]. In Fig. 4.1 this information is presented for two age cohorts, population aged 25-34 and population aged 55-64. Some 28 OECD countries are included in the comparison. It is clear that, for the younger cohort, in about half the countries more than 75 percent have completed upper secondary education and, in all but a small minority of countries, the 60 percent level is exceeded. The equivalent pattern for the older cohort is much more variable and in all countries the percentage is lower than that for younger people. In some cases, for example, Korea, Ireland, Belgium, Greece and Italy, the proportion is less than half of that for the younger cohort. The general picture suggests a process of convergence, with the vast majority of countries moving towards a state in which 80 or 90 percent of the population may be expected to have obtained some form of upper secondary school qualification.

For university-level education there are also sharp differences between the two age cohorts, and with one exception [Sweden], the percentage is higher among the younger group. Countries tend to fall into one of three categories: those where the percentage is relatively high for older people and is substantially higher for the younger cohort [e.g. Norway, United States and the Netherlands]; those where the percentage is low for both age groups [e.g. Czech Republic, Sweden, Denmark, Austria, Poland, Italy, Portugal and Turkey]; and those where the percentage for older people is low and that for younger people is much higher [e.g. Japan, Korea, New Zealand, France, Belgium, Ireland, Australia, United Kingdom and Spain]. It is not evident that there is a natural ceiling or that any general pattern is emerging.

Fig. 4.2 provides comparative data on literacy levels for a smaller sub-set of twelve countries. The definition used here is one provided by the International Adult Literacy Survey, which tested the knowledge and skills required to locate and use information contained in various formats such as official forms, timetables, maps and charts. Performance at Literacy Level 3 is generally considered desirable in order to avoid difficulties in coping with social and economic life in a modern democratic society (OECD 2000, p. 29).

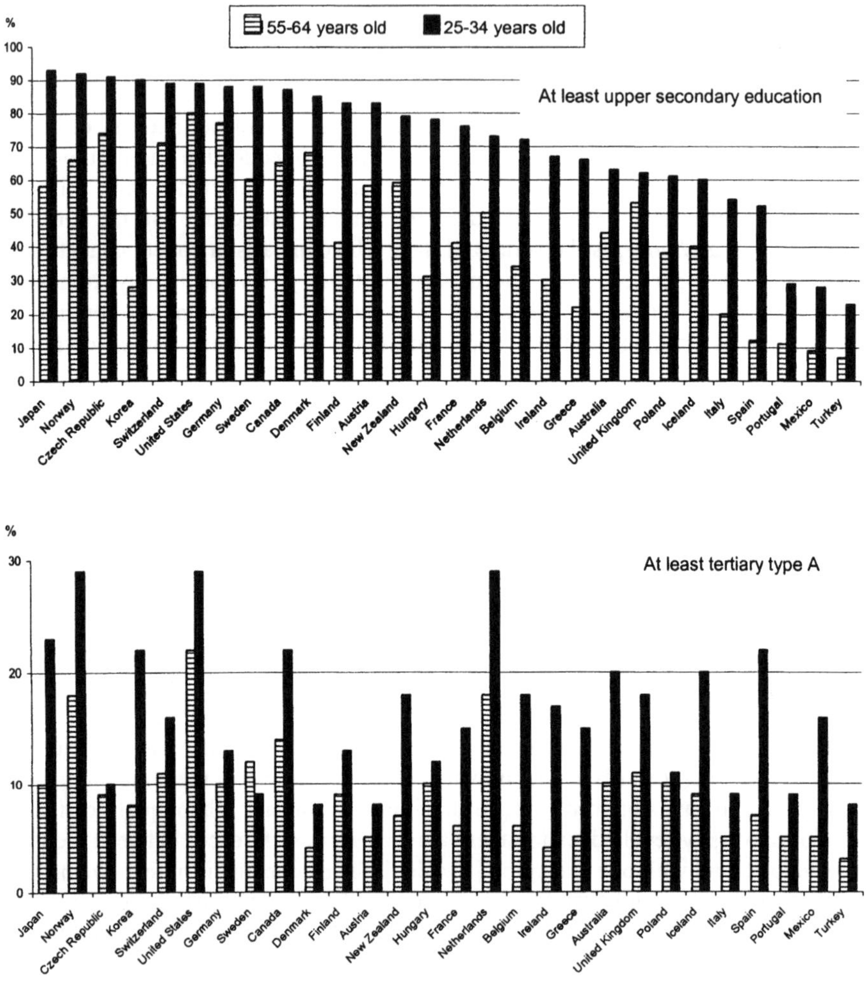

Source: OECD (2000, p. 28)

Notes: Year of reference: 1997. The level of educational attainment of tertiary-type A includes tertiary-type B.

Fig. 4.1 Percentage of the population that has attained at least upper secondary/tertiary type A level of education, by age group (1998)

The comparison in this case is between the 16-25 years age cohort, who are either in education or may recently have completed their education, and the 46-55 years ago cohort who are likely to have gained a considerable amount of experience since their schooling ended. In all but one country [United States], the literacy level among the younger cohort is higher, sometimes appreciably so [e.g. Netherlands, Belgium, Canada, Switzerland, Ireland and Poland]. There are signs of convergence among countries for the younger cohort, at between 60 and 75 percent.

Fig. 4.3 compares growth in the size of the university [tertiary-level] student population for a group of 25 countries. Comparisons are made between 1990, where the index is set for all countries at 100, and 1997. Change in student enrolment is expressed as a whole and in terms of two contributory factors, changing population size [more or less people in the relevant age cohort] and changing enrolment [participation] rates. The demographic component emerges as far less important than the change in enrolment rates. Several countries have experienced a dramatic proportional increase in their student population [e.g. Portugal, Poland and the United Kingdom], although it should be borne in mind that the starting period in some cases was very low, as Fig. 4.1 confirms. Comparison between Fig. 4.1 and Fig. 4.3 suggests that the sudden growth is quite recent in origin in some countries [e.g. Poland and Czech Republic]. A further feature that is worth noting is the relatively static student population level for countries where enrolment has historically been high [e.g. United States, Netherlands and Canada].

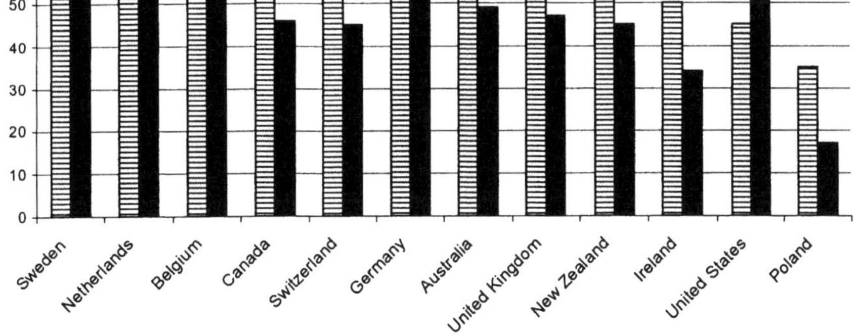

Source: OECD (2000, p. 31)

Fig. 4.2 Percentage of the population scoring at IALS literacy levels 3 or higher on the document scale by age group (1994-1995)

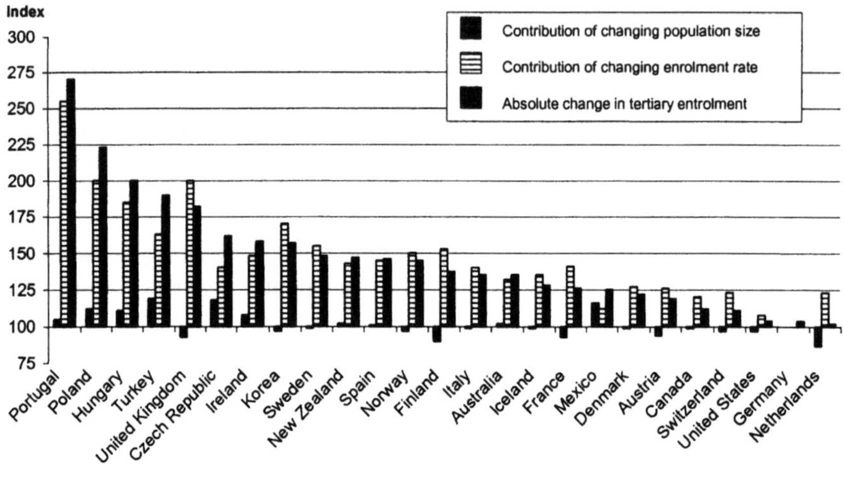

Source: OECD (2000, p. 154)

Fig. 4.3 Index of change in the number of students at the tertiary level between 1990 and 1997, and contribution of demographic changes and changing enrolment rates to the change in tertiary enrolment (1990=100)

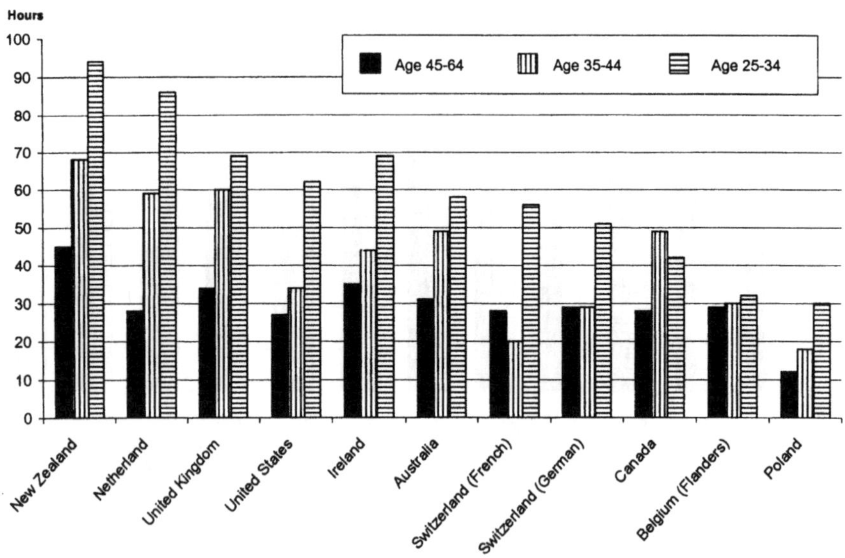

Source: OECD (2000, p. 195)

Fig. 4.4 Mean number of hours of job-related continuing education and training per employee and year, by age (1994-1995)

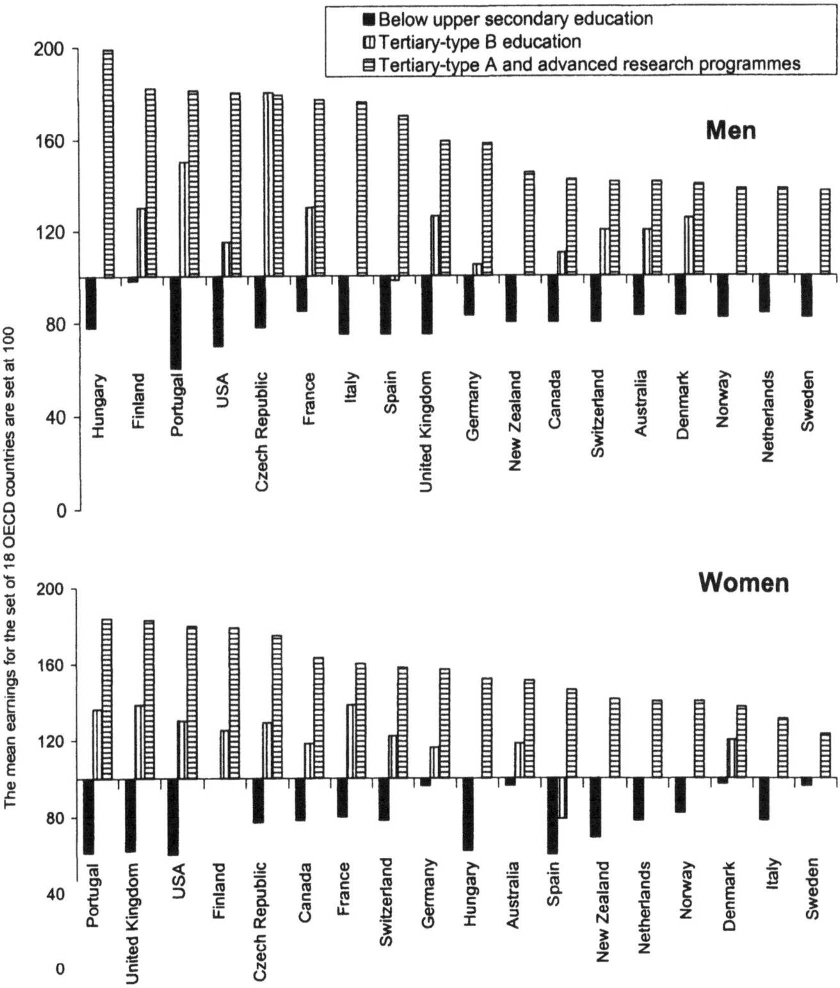

Source: OECD (2000, p. 293)

Fig. 4.5 Relative earnings of 25-64 year-olds with income from employment by level of education attainment and gender

The concept of lifelong learning has been heavily promoted in many OECD countries and is based on the notion that workers need to continue to learn new skills and practices, to remain competitive in the job market, and that firms need to instil a learning culture if they themselves are to compete effectively. Fig. 4.4 shows comparative data for ten countries and distinguishes between three age cohorts: 25-34, 35-44 and 45-64.

The data measure the average of time spent on job-related continuing education and training per worker per year. It is very clear from these data that younger workers are far more likely to engage in lifelong learning than older workers. The principle of lifelong learning is best established in New Zealand and the Netherlands and substantially less important in Canada, Belgium and Poland. The figures for the 25-34 years cohort suggests much higher priority is being given to ensuring that this group of workers receives job-related education and training than for their older colleagues.

The final comparison looks at the relative earnings of workers by level of educational attainment and gender for 18 countries (see Fig. 4.5). Three educational attainment levels are distinguished: below upper-secondary level education; tertiary-type B education [sub-undergraduate degree][3] and tertiary-type A [undergraduate degree and above]. The figures refer to 25-64 year olds and earnings for those completing upper secondary [and post-secondary non-tertiary] education are set at an index value of 100. The figures show clearly the positive impact of university education across all countries, but particularly in Hungary, Finland, Portugal, the United States and Czech Republic. The picture varies, however, between men and women. The financial benefits are generally greater for men than for women, with the notable exception of the United Kingdom, Canada, Australia and the Netherlands where graduate women can expect a bigger uplift in earnings than their male counterparts. The earnings of those who did not complete upper secondary level education lag behind those of their better qualified colleagues most in Portugal, the United Kingdom and the United States.

The overall impression gained from these international comparisons is that there is no consistent pattern in which a particular group of countries emerges as leading, or lagging behind, on all of the indicators. In all cases there is some scope for developing specific aspects of human resources.

4.3 The Learning Economy, the Learning Society and Learning Regions

Much of the recent discussion about the role of human resources in regional development draws upon what is known as the learning region paradigm (Boekema et al., 2000). The idea of a learning region can be seen as an extension of work by Lundvall and others (Lundvall, 1992; Lundvall and Johnson, 1994) on national systems of innovation which led to the notion of a 'learning economy'. Economic growth, at national and regional levels, is viewed as dependent on innovation, and innovation in turn relies on the creation, dissemination and application of knowledge (Schumpeter, 1926), usually referred to as learning. According to Lundvall, a learning economy is one where the success of individuals, firms and regions reflects the capability to learn [and forget old practices]; where change is rapid and old skills become out-of-date and new skills are in demand; where learning includes the building of competencies, not just increased access to information; where learning is going on in all parts of society, not just in high-tech sectors; and where net job creation is in knowledge-intensive sectors [high levels of Research and Development, a high proportion of university graduates, with the result that the unskilled find it harder to compete for jobs] (Goddard, 1999).

It is helpful to relate the idea of a learning economy to the process of globalisation and competition. Maskell (1998) has suggested that the globalisation of the economy has led to the creation of a new knowledge-based economy as a result of a process he describes as 'ubiquitification'. Many of the competitive advantages previously enjoyed by particular countries and regions have declined because the knowledge and technologies on which they depended are now available universally and the successful practices of firms have been replicated elsewhere. Firms in low-cost countries and regions have benefited from this process, in so doing threatening the survival of western firms. What used to be localised inputs are now ubiquitous because of the globalisation of the economy. They are no longer a source of competitive advantage.

Boekema et al. (2000) draw a parallel between these observations and Porter's (1990) distinction between higher-order and lower-order competitive advantages. Lower-order advantages, such as labour costs, the availability of raw materials, and the use of technologies, are easy for competitors to copy or compensate for, especially in an international economy with negligible transport costs and increasingly global markets. '*Higher-order advantages, such as proprietary process technology, product differentiation based on unique products or services, brand reputation ... customer relationships protected by high customer costs of switching vendors, are more durable*' (Porter, 1990, p. 50). In Porter's view, higher-order advantages have certain features that make them more durable: they require more advanced skills and capabilities to achieve them; they depend on '*sustained and cumulative investments in physical facilities and specialised and often risky learning, research and development, or marketing*'; and a combination of larger investments with a superior performance of the activities involved (Porter, 1990, p. 50).

These higher-order advantages are less prone to ubiquitification and require more effort to put in place. According to Boekema et al. (2000) firms must invest '*in the creation of more stable advantages, an activity which is closely related to the creation of enhanced knowledge used for the development of firm-specific competences ... the creation of competitive advantage is an act of innovation through which firms develop new and better ways of competing*' (Boekema et al., 2000, p. 6).

Several different kinds of knowledge are found in the learning economy. The first of these, 'know what', refers to facts and information. The second, 'know why', includes principles and laws necessary to reduce trial and error. The third, 'know how', describes the skills and capability to do something, the kind of knowledge that is usually acquired 'on the job'. Finally, there is 'know who', information about who knows how to do what and the social capability to build up learning relationships with specific groups and individuals and to draw upon their expertise. Each of these forms of knowledge is transmitted through different channels, so that for 'know what' and 'know why' the most important source will be schools and universities, whereas 'know how' depends on practical experience obtained by 'tacit' learning in the workplace and by network relationships with industrial or commercial partners, and 'know who' is learned from social interaction with a range of professional contacts, customers, suppliers and other organisations and agencies (Goddard, 1999).

Knowledge can be grouped into tacit knowledge and codified knowledge (Nonaka and Takeuchi, 1995). The former is difficult to transfer as it tends to refer to a specific human and social context. According to Nonaka and Takeuchi, these two types of knowledge are complementary. The creation of knowledge involves converting tacit knowledge into codified knowledge and vice versa. New knowledge which an organisation may want to absorb is often tacit and must first be codified to make it transferable. People in the organisation have to learn to use this knowledge which means that the new knowledge must become a part of their, and their organisation's, pool of tacit knowledge. Boekema et al. (2000) suggests that codified and tacit knowledge are constantly in interaction with each other, thereby creating 'a dynamic spiral of knowledge conversion leading to innovation' (Boekema et al., 2000, p. 7).

Goddard identifies network knowledge, related to 'know how', as worthy of special attention. Such knowledge is a hybrid of knowledge that is neither completely public nor completely private. '*It depends on trust, not the market, and is characterised by such considerations as reliability, honesty, cooperation and a sense of duties to others. Network knowledge refers not only to the skills of individuals but the transfer of knowledge from one group to another to form learning systems – the institutional infrastructure of public and private partnerships*' (Goddard, 1999, p. 40). Network knowledge is heavily dependent on interpersonal relations which means that it is often best developed within a particular region. Codified knowledge, on the other hand, can easily be transferred on a global scale, for example by using the internet.

Florida (1995) argues that, in order to be effective in an increasingly borderless global economy, regions must seek to behave like a knowledge intensive firm, and should therefore focus on '*continuous improvement, new ideas, knowledge creation, and organisational learning. Regions must adopt the principles of knowledge creation and continuous learning; they must in effect become learning regions*' (Florida, 1995, p. 32).

One consequence of moving towards a knowledge economy is that it places an obligation on the working population to engage in lifelong learning. Opinions vary as to how much this can be left to the individual or encouraged by means of social support systems for learning. In a debate promoted by the Royal Society of Arts in Britain, Ball (1999) stressed individual responsibility, whereas Coffield (1999) placed emphasis on strengthening the social support systems for learning. Recent initiatives have involved policies for lifelong learning operating at three interconnected levels: at the micro level through the introduction of individual learning accounts, essentially vouchers that can be used by individuals to buy approved forms of training; at the meso level, by promoting improvements within firms or institutions, for example by the adoption of quality circles or registering for Investor in People status; and at the macro-scale, for example through the new University for Industry and through Foundation Degrees, national schemes for increasing the opportunities for workers to receive additional, work-relevant education and training. Coffield (1999) expresses some concern that the culture of lifelong learning may not be encouraged in much of British industry: '*The new skills of computing, communication, problem-solving and teamwork can be developed only at work. It has been estimated that about two-thirds of the British workforce do not have access to organisations which develop such skills, and inequalities of access cannot be remedied by individuals but need to be addressed by government and employers rather than individually*' (Coffield, 1999, p. 86). To judge from the evidence of Fig. 4.4, a similar comment may well be applicable in many other developed countries.

Although there is increasing awareness of the need to create learning regions and to develop appropriate policies, in many cases there is a considerable amount of work to be done to prepare regions for this role. As Goddard (1999) points out, in many countries universities have traditionally produced new graduates for a national labour market dominated by large employers, with little concern for the needs of SMEs or with graduate retention in local labour markets. In some cases regional agencies have been formed to promote graduate retention initiatives to upgrade the stock of higher-grade local skills, aimed specifically at SMEs, but the impact of such schemes is still quite small. Measuring the impact is not easy because the quality of labour market information is generally low. This means that little is known, for example, about the flow of students through higher education within local labour markets and how this affects the economic performance of regions. There is also a need for more work to establish future labour market trends, to determine what are the appropriate skills for a region.

4.4 The Role of Universities in Regional Development

The learning region paradigm suggests that a reconsideration of the role of universities in regional development is needed. In the past it would have been common to view this role as rather limited: most discussions would have been confined to a concern with the employment effects associated with staff and student spending in the local economy and to the issue of promoting technology transfer by creating spin-off companies and the establishment of science parks (Department of Education and Employment, 1998; Goddard, 1999).

In the last five years especially the scope of this role has widened. A recent statement from the North West Universities Association in Britain is typical of this broader perspective. Fig. 4.6 lists seven ways in which universities see themselves as contributing to regional development. Particularly worthy of note are the statements about lifelong learning, the environment, social and community development and cultural developments. To this list could be added a number of further roles: recruiting students from outside the region and placing them in local companies [the graduate retention role referred to in Section 4.3], providing strategic analysis and leadership within the local and regional community, providing a global knowledge base for SMEs and locally embedding global businesses by target training programmes and research links (Goddard, 1999). In the British context, the recent creation of a Higher Education Innovation Fund [HEIF] as a third leg [in addition to teaching and research income steams] of the Higher Education Funding Council for England [HEFCE] allocation of government funding for universities is indicative of attempts to change traditional attitudes about universities' linkages with the business sector, regionally and nationally.

1. Contributing to lifelong learning through increased access, continuing education and training for companies
2. Transfer of knowledge and technology to boost competitiveness and innovation of local businesses
3. The environment, including the built environment of the region
4. Tourism and conferences
5. Job creation and the recruitment of students with spending power
6. Social and community development including impacts on society
7. Cultural developments in increased audiences for live theatre, performing arts and a variety of museums and galleries

Source: North West Universities Association (2000, p. 2)

Fig. 4.6 The contribution of universities to regional development. Example: North West Universities Association [UK]

Felsenstein (1996) has developed a very useful analytical framework for considering the role of universities within their local region. This framework is shown in Fig. 4.7. It represents a university as an organisation that receives inputs in order to generate certain outputs. The inputs to the university come from households, government and business and are shown in Fig. 4.7 as a series of backward linkages or expenditure effects. Expenditure by the university influences the levels of income, consumption and employment among local households, it supports the tax base and consumes services of local government and it supports a certain volume of activity in local business.

The outputs of university activity can be viewed as human capital formation, the production of knowledge and the creation of an attractiveness value. These forward linkages are the so-called knowledge effects and tend, according to Felsenstein, to be more long-term in their impact than expenditure effects. The university generally raises the level of human capital locally and so raises the productivity of all labour in the region [since the skill composition of the labour force will affect the technology used]. The university, by increasing the level of human capital, will also have an effect on business location decisions; a skilled-labour pool may attract existing firms from other regions and can influence the rate of local new firm formation. However, both these effects will only be felt if some of the students attracted by the university stay on in the region after graduation [the graduate retention issue]. The university will generate knowledge, some of which will be sold to export markets and some will be used to help local firms become more competitive. The university's presence in a region creates a local non-market value which may increase the region's attractiveness to households and, in some cases, firms, by means of social and cultural events that might otherwise not be viable.

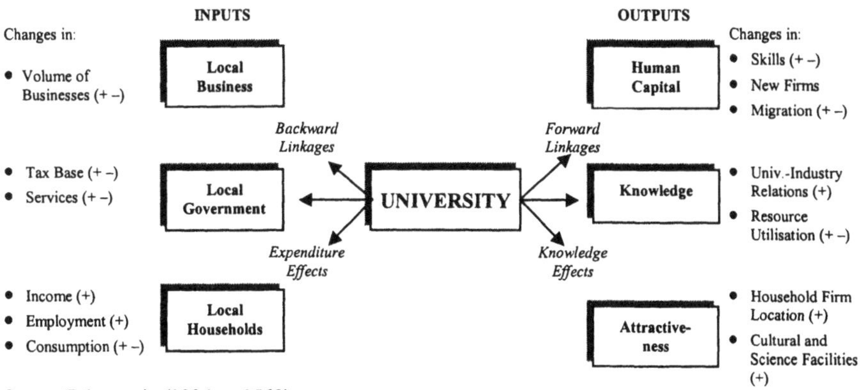

Souce: Felsenstein (1996, p. 1569)

Fig. 4.7 Metropolitan development impacts associated with a university [+ – positive and negative changes]

Felsenstein's framework nicely balances the consideration of expenditure effects – the subject of most university impact studies to date – with that of knowledge effects, much more important in the context of human resources and regional development.

4.5 Setting Learning Targets: A Regional Perspective

In this section the focus is upon measuring progress towards learning targets. These are targets in a number of human resource fields that have been set at the national level but applied to each of a number of regions. Experience is reviewed from England where the central government Department of Trade and Industry established a series of fourteen Regional Competitiveness Indicators in 1997. These indicators, which are updated twice a year, or more often if data are available, cover such topics as GRP [as a general measure of regional competitiveness], household disposable income, labour productivity, unemployment and, of particular relevance here, educational and vocational attainment and participation by firms in the government's Investors in People scheme which encourages lifelong learning that is employment-related.

These Regional Competitiveness Indicators [RCIs] make use of a number [but not all] of National Learning Targets set by the UK government for England in 1998. These targets refer to the levels that a series of indicators should reach by the year 2002. The indicators chosen for RCI treatment are the following:

National Targets for Young People:
- 85 percent of 19 year-olds will be qualified to at least NVQ [National Vocational Qualification] Level 2 or equivalent. [Here NVQ Level 2 refers to a basic craft vocational qualification; equivalent academic qualifications include five GCSE [General Certificate of Secondary Education] passes at grades A-C, examinations generally taken after five years of secondary school education].

National Targets for Adults:
- 28 percent of economically active adults will be qualified to at least NVQ Level 4 or equivalent. [Here NVQ Level 4 refers to a higher technician/ management vocational qualification; equivalent academic qualifications include sub-degree university qualifications].
- 50 percent of economically active adults will be qualified to at least NVQ Level 3 or equivalent. [Here NVQ Level 3 refers to an advanced craft/technician/supervisor vocational qualification; equivalent academic qualifications include GCE [General Certificate of Education] A Levels, examinations generally taken after seven years of secondary school education].

National Targets for Organisations:
- 45% of organisations with 50 or more employees will be recognised as Investors in People. [The IIP [Investor in People] scheme requires employers to ensure that everyone working for them understands how their job contributes to what the organisation wants to achieve, receives the training and development required to help them perform even better, takes time to plan for the training and development of their staff, and measures the return on the investment in training and development].

Table 4.1 Percentage of 19 year-olds with an NVQ Level 2 or 5 GCSEs at Grade C or above. Target: 85%, by 2002

	1998	1999	2000
North East	69	74	74
North West	73	76	78
Yorkshire and Humberside	75	76	72
East Midlands	70	73	73
West Midlands	67	73	70
Eastern	75	78	77
London	73	73	74
South East	79	77	79
South West	80	72	76
England	74	75	75

Note: Data refer to Autumn in each year

Source: Department of Trade and Industry (2001) *Regional Competitiveness Indicators*, Government Statistical Service, available on the Internet at www.dti.gov.uk

Table 4.2 Percentage of economically active adults qualified to at least NVQ Level 4 or equivalent. Target: 28%, by 2002

	1998	1999	2000
North East	21	21	22
North West	24	25	26
Yorkshire and Humberside	23	23	24
East Midlands	23	23	25
West Midlands	22	23	25
Eastern	24	24	24
London	35	35	36
South East	29	30	30
South West	26	27	27
England	26	27	28

Note: Data refer to Autumn in each year

Source: Department of Trade and Industry (2001) *Regional Competitiveness Indicators*, Government Statistical Service, available on the Internet at www.dti.gov.uk

Recent progress towards these targets is shown in Tables 4.1 to 4.4 for individual regions and for England as a whole. Table 4.1, which refers to a National Target for Young People shows that, in the time period under consideration, no region had reached the national target. In all cases, performance fell short of this target by at least five percentage points. Tables 4.2 and 4.3 refer to the National Targets for Adults. Table 4.2 shows that the target for NVQ Level 4 qualifications had been met for England as a whole by the year 2000 but that this was largely due to strong performance in London and the South East; other regions fell short of the target. Table 4.3 shows the position for NVQ 3 Level qualifications. Here again London and the South East are the leading regions and both had met the target by the year 2000. Other regions, such as the North East, the East Midlands, the West Midlands and the Eastern regions, still fell well short of the target by 2000. Table 4.4 refers to the National Target for Organisations. The 2002 target of 45 percent looks ambitious; by 2000 no region had exceeded 40%. In this case London is a lagging region and by 2000 it was some 17 percentage points short of the target.

Table 4.3 Percentage of economically active adults qualified to at least NVQ Level 3 or equivalent. Target: 50%, by 2002

	1998	1999	2000
North East	41	42	42
North West	45	46	48
Yorkshire and Humberside	42	45	45
East Midlands	43	43	44
West Midlands	40	42	44
Eastern	43	44	44
London	50	52	53
South East	49	50	50
South West	46	46	47
England	45	46	47

Note: Data refer to Autumn in each year
Source: Department of Trade and Industry (2001) *Regional Competitiveness Indicators,* Government Statistical Service, available on the Internet at www.dti.gov.uk

Table 4.4 Percentage of organisations [50+ employees] with investors in people recognition. Target:45%, by 2002

	1997	1998	1999	2000
North East	17	23	32	39
North West	15	20	27	34
Yorkshire and Humberside	12	19	24	34
East Midlands	13	19	26	33
West Midlands	15	21	26	31
Eastern	12	17	25	32
London	8	14	21	28
South East	14	19	28	33
South West	15	19	27	34
England	13	19	26	32

Note: Data refer to November in each year
Source: Department of Trade and Industry (2001) *Regional Competitiveness Indicators*, Government Statistical Service, available on the Internet at www.dti.gov.uk

The variation in regional performance on each of these indicators suggests that regional specific human resource development strategies are needed. In the next section, the purpose, form and content of such strategies are considered.

4.6 Human Resource Development and Regional Strategies

The production of *regional* competitiveness indicators and the setting of targets covering human resources topics implies that there are actions that can be taken at a regional level [as opposed to a national or a local level] to influence those targets. In the last five years there has been a considerable amount of progress in developing strategies and action plans for this purpose. In this section, an early example of a regional human resource development strategy is first considered, before examining the way in which the human resource dimension is embodied in a regional plan. The examples chosen are both from North West England.

One of the earliest human resource development strategies was *Learning for Earning*, a document produced by the North West Partnership in 1996. It was the first strategy of its kind to be produced for an English region. The strategy was intended to complement other strategies produced at European Union, central government and local levels. Its purpose was, first, to provide a framework for developments which can only be progressed effectively at a regional level, and to establish appropriate machinery and processes; and, secondly, to add value to initiatives of partners operating at a more local level, through greater cohesion of effort across the region, dissemination of best practice and learning from each other. The Strategy emphasised three aspects of human resources that challenged the conventional wisdom of the time: placing the workplace ahead of the campus; identifying the main problem facing the region as a shortfall in the demand for competencies [broadly defined] rather than in their supply; and underlining the importance of all forms of learning, not simply education and training. The Strategy's three objectives stressed competitiveness [developing relevant skills and competences in the region's workforce], employability [encouraging all workers to participate in learning and personal development], and social inclusion [encouraging and supporting the development of the least advantaged members of the region's workforce] (North West Partnership, 1996, p. i-iv).

The Strategy set out a number of regional targets for young people and for adults, along similar lines to the national targets that were reviewed in Section 5. It made a special point of identifying one target to which no indicator had yet been attached: 'On entering working life, young people should have the capacity and will to continue learning' (North West Partnership, 1996, p. v). This was seen as the single most important target: 'This is learning competence, the combined capacity and will to learn. For far too many of the region's young people this competence is still-born on entry to the working population; for many of the region's existing workforce it needs revival. The capacity to deal with unforeseen challenges, to take and make opportunities to acquire new knowledge and skills, to adapt and be flexible is the central point of the entire Strategy' (North West Partnership, 1996, p. iv).

Implementation of the Strategy was by means of a series of demand measures and supply measures, which are set out in Fig. 4.8. The demand measures were intended to provide partner organisations operating at a subregional level with a number of proposals to increase demand, whereas the supply measures were intended to refocus and raise the effectiveness with which providers of instructional services assist individual or collective learning (North West Partnership, 1996, p. vi).

Demand Measures	
DM1	Better information on jobs with learning opportunities
DM2	Better use of information on education and training opportunities
DM3	Better access to these opportunities
DM4	Employer support for individual self-investment
DM5	Worktime flexibility to improve use of competences
DM6	Better earnings incentives to learn
DM7	Shift to higher value-added goods and services
DM8	More high value-added goods and services
DM9	Promoting employer policies to increase skill development and use
DM10	Higher rate of net new business formation
DM11	Enhancing employer demand by reducing competence mismatch
DM12	Monitoring and updating of regional HRD strategy
DM13	Greater understanding of workplace HRD needs
DM14	Better access for the disadvantaged to learning opportunities
DM15	More effective technology transfer through training
Supply Measures	
SM1	Increased sensitivity of supply to circumstances and characteristics of individuals
SM2	More effective provision for the disabled
SM3	A strong and wider foundation for younger people for subsequent training
SM4	Achieving densities of qualified and highly qualified manpower comparable to competitors
SM5	Increasing use of participative learning methods in higher education
SM6	Increasing sensitivity of suppliers to needs of small and medium-sized enterprises

Source: North West Partnership (1996, p. vii)

Fig. 4.8 Learning for earning: The North West Human Resource Development Strategy – Implementation measures

The Learning for Earning Strategy was launched in 1996 at a time of institutional change within English regions. Its impact was largely in exposing human resources issues at the regional level for the first time. Its proposals for annually-updated action plans and regular monitoring of targets were never implemented fully and within two years of its publication, proposals were announced for a new system of regional administration involving representative Regional Assemblies and Regional Development Agencies. Significantly, one of the North West Development Agency's first tasks, when it was established in April 1999, was to commission the preparation of a Regional Learning and Skills

Strategy. *The Right Angle on Skills: North West England*, published in late 1999, follows many of the same principles as *Learning for Earning*, defines a more elaborate system of performance targets and provides clearer guidance by which agencies and at what levels action is needed.

The second example looks at the experience of Merseyside, an older industrialised subregion within the North West of England. It has a high level of unemployment and GRP per head is relatively low. In 1993 the Merseyside region was designated an Objective One region, making it eligible for the highest level of European Structural Funds. After receiving a first phase of Objective One funding (1994-1999), Merseyside has now begun to implement a second phase, covering a seven-year period, 2000-2006. In this programme and in all other programmes of the same kind, European Structural Funds are matched by substantial injections of public and private sector funding. The programme is guided by a Single Programming Document that covers the entire range of European Social Fund [ESF] and European Regional Development Fund [ERDF] activity.

The Single Programming Document includes a substantial human resource development component. Of particular interest is the fact that the structure and content of this component is to a large extent determined by higher-level human resource development strategies at European Union and national [UK] government levels. Three policy documents are especially important:

- The *European Employment Strategy* [and its four pillars: improving employability; developing entrepreneurship; encouraging adaptability of businesses and their employees; and strengthening equal opportunities for women and men, enshrined in the Amsterdam Treaty];
- *The National UK Employment Action Plan* – the national response to these European guidelines;
- The *UK Policy Frame of Reference* for promoting employability and developing human resources. This provides the overall structure for operating ESF in all parts of the UK. It links closely with the European Employment Strategy and with the National Employment Action Plan. Five priorities are identified: active labour market policies, equal opportunities for all and promoting social inclusion; lifelong learning; adaptability and entrepreneurship; and improving participation in the labour force.

The Single Programming Document is divided into four priorities: developing business, developing people, developing locations and developing pathways communities. The main emphasis on human resources is found in Priority 2 which contains a total of six measures, five of which relate directly to the policy fields in the UK Frame of Reference and are funded through ESF, and the sixth an ERDF-funded measure. Some of the measures also appear in the Developing Pathways Communities priority, targeted on particularly deprived neighbourhoods. The content of each of these measures is summarised in Fig. 4.9.

Action for Employment is a substantial measure within this priority, and uses a people-centred approach to move people towards sustainable employment. It is divided into two sub-sections, the first is for preventative measures, for those identified as at risk of becoming long-term unemployed, and covers not only those recently unemployed, but also young people new to the world of work and those still at school, provided that they are seen to be at risk of exclusion from the labour market. The second section is for those who have already experienced substantial unemployment, and will require concerted effort and support to bring them to the stage where they can gain employment. Strategies will be developed by District Partnerships, with the active participation of businesses, the voluntary sector, trade unions, representatives of excluded groups, and lined with other programmes.

Promoting social inclusion for access to the labour market is the other major measure within this priority, and aims to tackle the deep-seated problems of social and economic exclusion on Merseyside. While Priority Four is concerned with area-based regeneration [Pathways Communities] and exclusion based around geographical issues, this measure covers exclusion wherever it is found on Merseyside, but focussed on the needs of the individual, and the barriers which need to be addressed which enable people from excluded groups to participate fully in society. This measure will include special arrangements for access to small grants by voluntary and community organisations. These special arrangements will be based upon the selection of intermediary bodies to disburse grants in accordance with the Programme.

Lifelong learning links with the overall strategy for the UK and is expected to promote and add value to the core activities of the Learning and Skills Council for Merseyside and Halton. It will be primarily concerned with activities to achieve major changes in culture, systems and structures for the delivery of education, training and counselling, but will also support activities amongst those of school and working age with the aim of increasing attainment and skill levels towards National Learning Targets. This activity will complement measures elsewhere in the Programme to enhance skills as part of improving business competitiveness and improving the employability of those seeking work. The strategy will complement and add value to mainstream UK learning initiatives. Specific strategies will be developed to support greater equality of opportunity in the labour market for defined priority groups in particular in respect of occupations and industries where they are currently under-represented.

Job creation and entrepreneurship. This measure concentrates on development of the skills and activities people require to start to operate small businesses. In particular, this will be community based enterprises, the social economy and Intermediate Labour Market activities.

Positive Action for women. Equality between men and women is mainstreamed throughout this programme. In addition, this measure is specifically aimed at affecting the gender pay gap and horizontal and vertical segregation on the basis of gender, by addressing women's access to and participation in the labour market, and women's professional development.

Facilitating access to the labour market. The ERDF funded part of this Priority is designed to complement and enhance the ESF activities. It will strengthen professional, institutional and community capacity for delivering education and training, social inclusion and equal opportunities, and will support better monitoring and evaluation.

Source: Merseyside Strategic Partnership (2000)

Fig. 4.9 The Merseyside Single Programming Document 2000-2006 Measures contained in Priority 2 – Developing people

The Merseyside document is typical of the way the European Union disburses its Social Fund to support the development of human resources. To a large extent the strategic direction of the regional plan was set by the higher-level policy statements; it is only the means of implementation which is left to the discretion of the regional partnership. What is also interesting is the degree to which the approach adopted is standardised within a particular country; in this case the UK Policy Frame of Reference serves the purpose of ensuring that common issues and concepts are followed by all regions eligible for Structural Funds support.

4.7 Conclusions

This chapter has reviewed current thinking about human resource development as an element of regional planning. It has presented evidence that concepts and paradigms first developed as part of national strategy are now being translated into a regional context. The knowledge economy, now a cornerstone of government policy in many developed countries, finds expression as the learning region. Such a region is one in which all sections of industry and commerce are responsive to the need for continuous improvement, innovation and knowledge creation.

The chapter has also considered the contribution that universities can make to regional development. Traditionally the emphasis has been upon expenditure effects involving local households, local government and local businesses [backward linkages], but more recently attention has shifted to knowledge effects, in terms of enhanced skills, new firm formation, and in-migration of workers and firms [forward linkages]. It is these impacts that are important in relation to human resource development. A much fuller role is now seen for universities in their local regions.

The use of performance indicators and the setting of targets for human resources are another theme that has been explored. The chapter has examined international variations among developed countries on a number of key indicators, including the proportions of the workforce holding upper secondary school and university qualifications, literacy levels, university education participation rates and involvement in continuing professional development. The development of regional competitiveness indicators measuring the attainment of vocational and academic qualifications by young people and adults has been considered, as a means of monitoring the achievement, at regional level, of national education and training targets.

Finally, two examples of regional human resource development strategies have been examined. These examples showed the importance of workplace development of human resources, the need to stimulate demand for work-related education and training and the steps that must be taken to overcome social exclusion. The second of the two examples demonstrated the linkages between different tiers of human development strategy, at European Union, national and regional levels.

Many of the developments presented here in terms of concepts and practice are very recent in origin and to some extent, much of the experience is the product of experiments in policy which have still to be evaluated fully. Without doubt, however, human resource development in one form or another, is now accepted as an essential element of regional development strategy.

Endnotes

1 Upper secondary level of education generally refers to schooling that is either preparatory, i.e. preparing students for tertiary education, or terminal, i.e. preparing students for entry directly into working life. Upper secondary schooling takes place after about three years of lower secondary education and tends to last four or five years (OECD 2000, p. 372).

2 Tertiary Type A degree programmes are largely theory-based and are designed to provide sufficient qualifications for entry to advanced research programmes and professions. They take a minimum of three years full-time study to complete (OECD 2000, p. 372).

3 Tertiary type-B programmes are typically shorter than those of tertiary type-A and focus on practical, technical or occupational skills for direct entry into the labour market. They have a minimum duration of two years full-time equivalent at the tertiary level (OECD 2000, p. 372).

References

Ball C. (1999): Our Learning Society, *Journal of the Royal Society of Arts* 4 (4), 83-84

Batey P.W.J. and Friedrich P. (eds.) (2000): *Regional Competition*, Springer, Berlin. Heidelberg, New York

Boekema F., Morgan K., Bakkers S. and Rutten R. (eds.) (2000): *Knowledge, Innovation and Economic Growth: The Theory and Practice of Learning Regions,* Edward Elgar, Cheltenham

Coffield F. (1999): Our Learning Society, *Journal of the Royal Society of Arts* 4 (4), 84-87

Department of Education and Employment (1998): *Universities and Economic Development*, DFEE Higher Education and Employment Division, Sheffield

Department of Trade and Industry (2001): *Regional Competitiveness Indicators*, Government Statistical Service, available on the internet at www.dti.gov.uk

Felsenstein D. (1996): The University in the Metropolitan Arena: Impacts and Public Policy Implications, *Urban Studies* 33 (9), 1565-1580

Florida R (1995): Towards the Learning Region, *Futures* 27 (5), 527-536

Goddard J. (1999): How Universities Can Thrive Locally in a Global Economy. In: Gray H. (ed.) *Universities and the Creation of Wealth*, Open University Press, Buckingham, pp. 36-46

Lundvall B.-Å. (1992): *National Systems of Innovation: Towards a Theory of Innovative Learning*, Pinter, London

Lundvall B.- Å. and Johnson B. (1994): The Learning Economy, *Journal of Industry Studies* 2, 23-42

Maskell R. (1998): *Competitiveness, Localised Learning and Regional Development: Specialisation and Prosperity in Small Open Economies*, Routledge, London

Merseyside Strategic Partnership (2000): *Merseyside Objective 1 Programme 2000-2006 Single Programming Document*, Government Office for the North West, Liverpool

Nonaka I. and Takeuchi H. (1995): *The Knowledge Creating Company: How Japanese Companies Create the Dynamics of Innovation*, Oxford University Press, New York, Oxford

North West Development Agency (1999): *The Right Angle on Skills: North West England*, NWDA, Warrington

North West Partnership (1996): *Learning for Earning: A Human Resource Development Strategy for North West England*, North West Partnership, Wigan

North West Universities Association (2000): *Building on Success*, NWUA, Warrington

OECD (2000): *Education at a Glance: OECD Indicators*, Organisation for Economic Cooperation and Development, Paris

Porter M. (1990): *The Competitive Advantage of Nations*, Macmillan, London

Schumpeter J. (1926): *Theorie der wirtschaftlichen Entwicklung: Eine Untersuchung über Unternehmensgewinn, Kapital, Kredit, Zins und den Konjunkturzyklus*, Duncker and Humbolt, Munich, Leipzig

UK Government (ed.) (1998): *Our Competitive Future: Building the Knowledge Driven Economy*, The Stationery Office, London

PART B: Competitiveness and Cooperation

5 The Role of Public Institutions in Regional Competition

Peter Friedrich and Xiao Feng
University of the Federal Armed Forces, Department of Economics and Organisational Sciences

5.1 Introduction

Historically, the notion of competitiveness has been associated with a relatively specific set of ideas and policies, in which public institutions are not given a central role. These earlier approaches are still the ones most familiar to many regional scientists. This chapter is focused on the role of public institutions in regional competition and begins by characterising regional competition and the role of public institutions [see Section 5.2]. Particular attention is paid to the issue of specifying regional competition among public institutions and a variety of models of political public regional competition are presented. Section 5.3 describes basic approaches. Section 5.4 moves to the case of political regional competition within an administrative sector, while Section 5.5 considers regional competition within a federation. Political fiscal regional competition and political competition of municipalities and public firms are discussed in Section 5.6. Section 5.7 concludes with a summary and a brief outlook.

5.2 Regional Competition by Public Institutions

Public institutions may be defined as public administrations or enterprises. To achieve their public goals they act autonomously according to the stipulations laid down in private or public laws. Public firms can be used as instruments of regional competition (Friedrich and Feng, 2000). This may happen, for example, in a framework of private/public partnership.

Regional competition can be characterised by its structure, processes, and results (Buhr and Friedrich 1978; Batey and Friedrich, 2000). The theoretical reflections on this matter are usually based on theoretical and methodological approaches. The methods applied refer to algorithms intended to derive solutions or to evolve empirical findings, to estimate relevant parameters or to gather empirical data and to investigate empirical observations. Policy oriented research deals with the context of regional competition, the objectives, instruments and evaluations. Fig. 5.1 makes a modest attempt to summarise and categorise the variety of approaches available. Single public offices, public firms and the jurisdictions they belong to, may act as public competitors in regional competition. They may also act as a single competitor within the region. Then microeconomic competition prevails. If they are viewed as a group, or if a public office acts as a representative of other economic units of a region, we are confronted with macroeconomic competition. Economic units act either in isolation or in cooperation in the course of regional competition. The aims of the economic units, may be either regionally or nationally oriented. Apart from horizontal competition among regions and the related economic units, vertical competition also exists. This is most often found between governments at various levels in the European Union and in member states which form federations. Regional competition can be described by the identification of regional competitors, the basic competition relationships, the instruments of competition, the rules and market forms of competition, together with the goals and results of competitors.

Competitors: Public offices, public agencies or public firms may act as representatives for a region in regional competition, thus becoming *competitors*. This can happen in three ways:

- They can directly represent a government or jurisdiction.
- Some public offices, agencies and public firms compete directly with other public or private economic units belonging to other regions. Depending on the field of regional competition, various public institutions are involved. Ministries of finance are actors in financial competition or tax competition at the national, state and local levels. In a few countries they even act at a municipal level. Public local development firms are sometimes responsible for city renovation. In cultural affairs, specific ministries or departments for cultural affairs (Funck, 2000), for science or education, and public firms such as theatres, etc. compete. In infrastructure competition the ministries of transport, traffic, communications, postal services, construction, regional development and city development are the main actors and competitors at the national and state government level. The departments for city planning and development, the departments of construction etc., and public firms compete at the local level. Public infrastructure utilities provide services, such as energy, water, irrigation etc., which regions use as competition instruments,. In the

sectors of health and social services, ministries, health departments and social departments play a role, along with public insurance companies, and public firms such as hospitals. Jurisdictions and their ministries of internal affairs and justice also compete by the provision of security services, and the control and regulation of services. A main field of regional competition is represented by business promotion. Ministries of economics, development agencies and public banks play a major role at national and state levels (Lindemann, 1999). At the EU level, the respective commissioner's office is responsible. To strengthen the position of regions in the various sectors, public enterprises are currently being restructured to build up clusters and to compete with private and public firms in nearly all sectors.

- Some departments and offices are responsible for raising funds in order to strengthen the position of a jurisdiction in regional competition. These can be public offices or specific companies whose purpose is to finance projects, or public banks and public firms such as casinos, energy supply firms, zoos, wineries, etc.

Basic competition relationships: Public offices *compete horizontally* although they may provide their services free of charge. They have service and procurement areas which may be determined by time related efforts, accessibility, preferences for the services provided, factor prises, etc. Projects regarding infrastructures involving the territories of several jurisdictions lead to horizontal coordinative negotiations. Bargaining can be influenced by complementary relationships among the competitors (Hewings and Sonis, 2000) that may limit the scope of action for horizontal bargaining among municipalities. Because of legal restrictions compensations are not easy to organise. In rare cases, common firms and institutions of private and public law will be established, where internal mutual compensations are possible. Sometimes the traditional horizontal market type competition takes place, if the services and goods carry a fee. Fig. 5.2 serves to identify and exemplify competitive relationships among regional competitors.

Public offices, but also agencies and public firms, are integrated into *vertical and hierarchical coordination*. This coordination may connect an EU commission office with a Federal bureau, a state office or even a municipal office. In such coordination structures several jurisdictions are involved. Sometimes vertical competition takes place between public offices which belong to the same jurisdiction, e.g. the federation, the state or the same municipality. If administrations of one sector become active in several regions which compete horizontally, the horizontal competition can be influenced by actions and decisions of a jurisdiction of a higher level. Vertical competition can be carried out through hierarchical coordination procedures under conditions of public law by means of decrees etc. In a more balanced power framework among the jurisdictions [offices] vertical competition leads to agreements, contracts etc. in the form of laws governing the public or private sectors. Vertical conflicts may exist between the proprietor of a public firm and the management of the public firm (Friedrich and Feng, 2000). The outcome is a network of horizontal and vertical regional competition that is partly reflected in Fig. 5.2.

			Regional Competition		
			Structure	**Process**	**Results**
Theory	**Definition**		Identification of public regional competitors, fields of regional competition of public competitors	Horizontal and vertical competitive relations, complementarity, competitive behaviour and actions	Political success in terms of votes, positions, considerations in plans, attraction of financial means, provision of public services by public administrations, economic growth
	Mode of Operation		Role of public competitors to attain or develop the public and economic order	Involvement in public policy programs in order to influence processes in favour of its own region	Results of public policy programs favourable or unfavourable for the public regional competitor
	Modelling Approach	**Negotiation/ Market**	Market forms or forms of negotiations with the public regional competitors involved, models of life cycle of public production	Strategies chosen, game or negotiation solution approach, horizontal competition models, principal agent models in vertical competition	Pay offs, stipulations and conditions laid down in contracts; allocation of factors, goods, political positions, competence prices, votes, centrality, etc.
		Growth/ Foreign Trade	Growth models involving infrastructure and public decision makers, political trade cycle models, societal development models, models applying foreign trade approaches to public institutions, federalism models, Tiebout models	Growth models of public offices, budget games, models of public sector development, models of negotiations regarding intergovernmental fiscal relationships, models of Wagners law, models concerning public, private and shadow sectors	Allocation of public activities, budget distributions, service and procurement areas, volumes of production, growth of sectors, tax receipts etc.
		Agglomeration/ Location/ Cluster	Agglomeration of private and public activities and economic units, location theory for public offices, public facilities etc., models of private-public-partnership	Linkages with respect to private activities following public ones and vice versa, coalition formation theories, models of political process	Degree of concentration of public activities, public employment, financial revenues, grants, etc.; distribution of voters, location of public offices, forms of private public cooperation and of public cooperation, administrative complexes

Fig. 5.1a Role of public institutions in regional competition: Theory

		Regional Competition		
		Structure	**Process**	**Results**
Method	**Algorithms**	Median voter models, vote models, coalition models, one or several decision makers	Modelling decisions in different political stages, simulations, OR-models, migration models	Programming results, results from differenc and differential equations concerning the achievement of goals
	Model Estimation	Econometric models to determine parameters in relation to public activities	Formulation of reaction functions: specification of decision makers behaviour, revealed preference analysis	Determination of effects and impacts as results of modelling, particular and general solutions of dynamic processes
	Empirical Observation	Methods for collecting statistical data on public institutions, interviews	Description of decision processes, historical analysis, investigation of sources, laws treaties, protocols, events	Description of results of distribution of locations, number of inhabitants, voters, density of public activities, public receipts, grants, etc.
Policy	**Context**	Economic order with priority to public or private sector; social market economy; local, national or international welfare orientation; role of public institutions on background of individualistic materialism; socialist, national-socialist and Marxist socialism, of Christianity, Islam, Buddhism, Hinduism, political domination of regions	Control and development by public intervention, public production, public firm, public responsibility for economic and social survival, privatisation, nationalisation, socialisation, public business promotion, protection	Regional level of welfare, sector shares and their changes, degree of autonnomy, degree of international market integration etc.
	Objectives	Public goals concerning society, state, economic situation, individual public offices, administrative goals, political goals sector goals, hierarchy of goals	Formation of goals, negotiation of goals, sources of goals such as laws, goals determined by legal, economic, cultural and other conditions, minimum requirements, measurement of goals	Achievement of goals and their acceptance, change in goals, losses, benefits, sanctions, rank of institutions
	Instrument	Public offices, private units, private-public institutions	Land use planning, formation of laws, sanctions, taxation, business promotion, granting, self-production, self-administration, measures of private law, measures of public law, negotiations	Effectiveness of measures, results of cooperation, impacts of measures, achievement of goals
	Evaluation	Isolated welfare measurement, welfare measurement, achievement of goals, goals in regional competition	Determination of evaluation methods, net-benefit analysis, cost effectiveness analysis, utility analysis, operalisation of goals, decision on goals	Evaluation results in terms of the chosen evaluation approach, consequences in evaluation results for changes in goals, actions, government formation, etc.

Fig. 5.1b Role of public institutions in regional competition: Method and policy

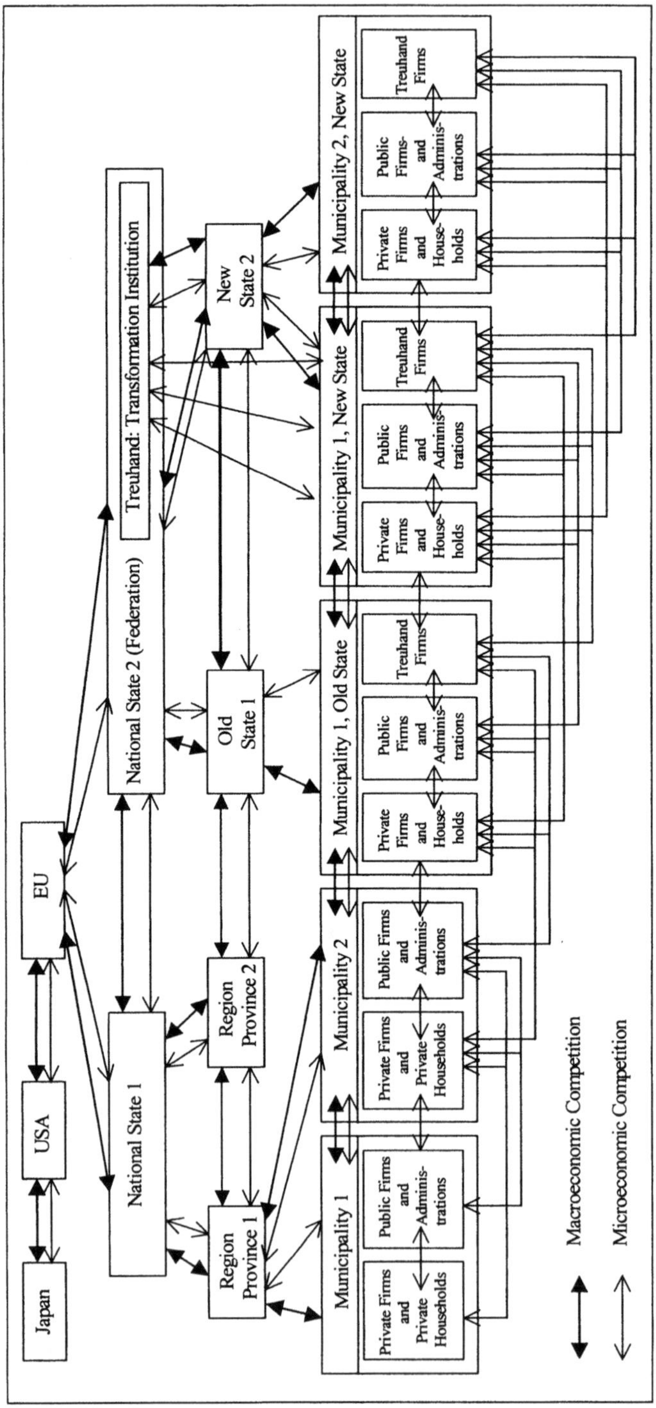

Fig. 5.2 Competitive relations in regional competition

There are also situations of horizontal, and to some extend vertical, regional competition between private or public economic units which belong to competing regions. Which kind of horizontal and vertical regional competition exists depends on the stage of development of the regions (Ritschl, 1931; Hirschman, 1958; Rostow, 1960; Hirsch, 1992; Parr, 2000; Poot, 2000), their possibilities of making use of innovations and of learning (Brezis and Krugman, 1997).

Instruments of competition are as manifold as are the horizontal and vertical conditions of the competitions to which they relate. Many instruments provided by rulings laid down in laws to be executed and by the stipulation of public laws have to be considered. Here, principles of equal treatment, lawfulness, avoidance of excessive measures, fulfilment of plans etc. become relevant. Competitors also intervene by means of regulations, sanctions, allowances, stipulations in plans, interpretation and specifications in laws and decrees, preparation of territorial and land use plans, as well as the development of special service programmes, or infrastructures. The quality of services, the provision of special services for education, vocational training, research, security, housing, the service times, as well as access to services, infrastructure, and political support, are instruments of regional competition in product policy. In the sphere of financial activities, the jurisdiction and its responsible administration can to some extent compete by subsidisation, measures of taxation, credit, guarantees, and the sale of property at favourable prices.

In the European Union the scopes of the actions taken by national states, but also by substates, provinces and national municipalities, are controlled by EU regulations and laws. This is also true for instruments of procurement, where jurisdictions can try to benefit the territorial industries when purchasing goods and services at home. The purchase of land for industrial purposes and real estate management in some municipalities plays an important role. Although procurement instruments are mainly those of horizontal competition, some instruments such as planning, co-determination, co-administration and controls specified in public law, are applied by competitors in vertical competition. In several sectors public enterprises are established to act as agents for principals which are responsible for a region. They are mostly active in sector specific regional competition. Depending on which economic unit is involved in horizontal or vertical regional competition and on the kind of the objects of the competition, different *market forms* result (Friedrich, 1976, 1992).

Market forms of regional competition differ largely according to the *kind of regional competition* in which the public institution is involved. When the top *public offices of a jurisdiction representing a region* are involved in tax, and infrastructure competition, and when there are vertical conflicts regarding financial means and regional planning, the instruments are mainly those used in negotiations and political struggles. Decisions on political moves and strategies must be made in councils, parliaments etc. Thus the *instruments of this kind of regional competition* are mostly found in the fields of zoning, the general fixing of tax rates if the authority has the right to use this as an instrument of competition, actions in infrastructure planning and general policies concerning the provision of public services such as education, health services, defence and security, waste management, and energy supply. An instrument of top level competition is the establishment of *public enterprises* which can be used to enforce the position of an authority in special fields of regional competition and as a general support for other economic units of the region.

In *vertical competition*, there are often bilateral monopolies when municipalities have to deal with their host state, or negotiate directly with Federal offices or an EU commission. Thy act within a chain of hierarchical offices of different authorities such as municipal, state, and federal institutions. A sequence of linked vertical bilateral monopolies exists. Sometimes only two jurisdictions are involved, where one jurisdiction deals with different public offices, for example a federal directorate and a federal ministry negotiating with a municipality. If the federal office and the state office have to decide together they may form a coalition to negotiate with the municipality, leading, for example, to a federation of municipalities, or a federation of states. Here the vertical competition is combined with horizontal competition when municipalities compete for financial means provided by the state, and infrastructure projects, etc. Several negotiation partners exist on the side of the lower level. The typical number of competitors of lower ranks is from two to ten. If the higher rank agency is the supplier, a demand restricted monopoly is created. If the high rank office is making the demand, a supply restricted monopsony results. Therefore, within the public sector, a tendency to market forms which do not comprise many market partners prevails. Moreover, no exchange of money for goods exists. Services, property rights, agreements, support etc. are changed or transferred in this kind of market.

Some *public offices are directly involved in horizontal sectoral competition favouring a region*, such as cultural institutions, institutions of higher education, offices controlling infrastructures, social institutions, sports, planning bodies, judiciary institutions, security agencies, armies, health administrations, etc. There is a sector specific set of instruments, which differ largely from the rules of public law, which apply and create possibilities of applying contracting, ruled by private law, delivery times, opening hours, qualities, information, service duties, etc.

In the case of infrastructure planning, cooperation may also occur. Here again competitors are sometimes neighbouring. Consequently, narrow oligopolies demand restricted oligopolies and supply restricted oligopsonies. More often than not, the competition is on the delivery side and the change of good against money is not of primary concern. Horizontal competition between metropolitan cities is often world wide, as can be seen by museums in New York, London, Paris, Berlin and Vienna. Here oligopolies prevail. Sometimes competition exists between cities of different rank, in which case oligopolies, or demand restricted oligopolies, are found. Competition can also exist between a central municipality and its hinterland communities with respect to various activities, such as holiday resorts amenities, supply of industrial sites, land use, schools etc. *Oligopolistic market forms* are found here once more. Monopolies are rare, e.g. courts, public attorneys, police offices. Many sector specific public facilities are managed by public firms.

Public firms are involved in sector specific horizontal competition such as sales and procurement. Oligopolies, oligopsonies, bilateral oligopsonies, supply restricted oligopsonies and demand restricted oligopolies prevail as a rule. Supply monopolies rarely occur in Europe. The course of regional competition, e.g. *municipal firm settlement competition,* is a process developed through several *phases* and related to *levels* of *decision making*. Towns compete within the framework of land use planning, as long as no individual projects are negotiated with investors. If individual investors offer settlement projects, towns compete through their business promotion agencies or companies (Lindemann, 1999; Friedrich and Lindemann, 2000; Müller and Kornmeier, 2000).

Very different *market forms* can be identified in competition among city governments and in competition among business promotion companies. Individual competitive relationships vary in different market phases, e.g., the experimental, expansion, maturity, stagnation and decline phases (Kühne, 1992). These phases overlap with respect to different firms, public offices and households. Possibilities of violent conflict arise if one region wants to subjugate economic units of another region. Public enterprises mostly behave oligopolistically and peacefully. They choose autonomous and oligopolistic strategies. Usually, the number of competitors in vertical competition is small. Bilateral monopolies or demand restricted monopolies dominate. Public firms tend to have conflicts with their public owners (Friedrich, 1992). Manifold *competition rules* and regulations have to be considered, since many commercial laws, e.g., antitrust laws, are binding for public firms as well. Public and constitutional laws affect horizontal and vertical competition among governments and partly among public enterprises. The system of legal rules concerning regional competition, – especially as the European Union, the Federation, state, and municipalities enact laws – is immense. Therefore, regional competition processes may only be explained with reference to the individual competitive measures of a specific region.

Institutions for improving the financial situation of a jurisdiction are sometimes public offices such as budget offices or special development agencies and funds. They often take the form of a public enterprise of public or private law such as casinos, public development firms, public banks, etc. Financial institutions, such as budget offices, are involved in horizontal competition including the sphere of sales of property, and in vertical competition for grants, subsidies, contributions, and the receipt of taxes. Often a demand restricted monopoly of the higher rank institution prevails. In so far as these institutions transact financially, the market is oligopolistic, or the relationships are bilaterally monopolistic.

According to a welfare function, public offices and firms should *contribute to the welfare* of the region. A welfare function, however, which refers to regional welfare is not available (Friedrich and Jutila, 2000). Therefore, a benefit cost analysis can be used to compensate for this lack. But the assessment of a benefit cost analysis which indicates an increase in so-called isolated welfare, involves insurmountable methodology.

Goals of regional competition: To perform a utility analysis, goal trees based on aim catalogues have to be formulated. Such aim catalogues for public firms, comprise regional goals of competition which are important in horizontal competition, especially with institutions representing a government or jurisdiction. Achievement of the indirect goals of competition, e.g. goals of public utilities, should improve the competitive situation of the recipient of these services (Thiemeyer, 1970; Eichhorn and Friedrich, 1976; Eichhorn, 1997). Goals for federal, state and municipal public offices refer to the general goals of society, such as peace, freedom, justice, security, and economic welfare, constitutional requirements for distribution. Basic goals of cities often depend upon the achievements of a higher rank in city hierarchy, as well as the growth of population. Some economic aims have to be considered such as full employment, growth, production goals, financial goals, political goals like maximisation of votes, posts, the defence and strengthening of political positions, as well as support of a policy or candidate. In addition to this, there are many thousands of goals in other sectors. With respect to vertical competition, the competition goals of public owners and of other governments play a role. These goals are often oriented towards increases in regional income, regional employment, regional production and improvement of the budget situation of the government responsible for a region. Horizontally and vertically, political goals are important (Gerken, 1999, p. 20). Political goals concern the maximisation of votes, the winning of a minimal majority in elections to form a coalition, and to meet the desires of the average voter[1]. These goals are the same for public enterprises.

Results of regional competition are illustrated by the degree of attainment of these goals, the achievement of competitive goals of public offices and firms, and the contribution of public offices and firms to the attainment of these goals. Models of political regional competition concerning the localisation of public offices include game theoretic approaches (Friedrich, 1976), as well as models to trace politically relevant location effects (Wonnemann, 1989), and models of transformation by sales, transfer and privatisation of publicly owned firms in regional competition (Feng, 1998). They help to identify the political results of regional competition. The outcome of political oriented competitive decisions may be identified in economic terms such as in tax competition, subsidy competition, fiscal competition, infrastructure competition, location competition, industrial competition, development competition or environmental competition. For the analysis of regional competition being considered by public firms, models of price theory (Krelle, 1976; Friedrich, 1969, 1977; Bös, 1986) as well as game theory (Lindemann, 1999), control theory (Putnoki, 1990), economic evolutionary approaches (Batey and Friedrich, 2000) and concepts of competition are studied. We refer to the economic and political results of regional competition.

5.3 Political Models of Regional Competition: Basic Approaches

When regional scientists discuss implications of concepts of regional organisation, measures and impacts of regional policy (Cheshire and Gordon, 1998) in the context of regional competition, they generally do not consider the influences of political factors which determine decisions in public institutions taken by political decision-makers[2]. In recent times, some researchers have considered the competition of jurisdictions within federal states more intensively. '*Functional, overlapping, competing jurisdictions*' [*FOCJ*] are task oriented jurisdictions, e.g., for the supply of water, that may overlap spatially and may compete for citizens or municipalities (Gerken, 1995; Frey, 1997, 1999). In this regard FOCJ represent a specific juridical form of jurisdiction for specified administrative purposes, rather than for specified functions comprising a whole administrative sector, such as the European Armed Forces, European Railways, European Air Control Jurisdiction etc. A FOCJ should have the right to tax and its government should be determined by a parliament made up of the FOCJ, and whose members are delegated by the members of the FOCJ. Members of a FOCJ may leave or join the FOCJ according to conditions stipulated by the FOCJ itself. A FOCJ may represent a competing region or it may hinder the group formation in favour of a region, because it overlaps with competing regions. In European countries such kinds of jurisdictions already exist, for example, as a special legal form for some types of public enterprises, associations, churches,. The discussion of FOCJ draws attention to administrative and political competition (Friedrich, 1977; Dohse, 2000). Schwager (2000) offers a game theory model to elaborate conditions for the formation of a FOCJ.

Particular emphasis will later be laid on

- Political regional competition within an administrative sector [see Section 5.4],
- Political regional competition within a Federation [see Section 5.5].
- Political fiscal regional competition and political competition of municipalities and public firms [see Section 5.6].

5.4 Political Regional Competition within an Administrative Sector

Processes carried out between public offices within an administrative sector that influence the abilities of a region to compete are often neglected, due to a lack of formalised theories influencing administrative public activities. There are a few exemptions (Niskanen, 1971; Eichhorn and Friedrich, 1976; Borcherding, 1977;

Friedrich 1984; Friedrich and Pfeilsticker, 1985; Renault and Winden, 1991). We consider a simplified three-*level administrative system* of a jurisdiction [see Fig. 5.3]. An administration of rank III [government, parliament] which – as part of the administrative and political system of the jurisdiction exists – has to fix the overall budget volume D. The rank III administration coordinates by budget assignment the rank II administrations, one of which is considered in the model. Middle-level rank II administrations coordinate low-level rank I administrations by allocating the budget D to them (D^I_1, D^I_2). Rank I administrations provide services X^I_1, X^I_2 for region 1 and region 2. Voters living in region 1 and in region 2 react positively to higher output, but negatively to financial burdens. According to Batey and Friedrich (2000) maximisation of the sum of votes for both regions yields the optimal budget with regard to the administration of rank III. Factor prices, evaluation of management in administrations of different levels and in different regions, the regional sensitivity of voters [expressed by the parameters g_{X1}, g_{X2} and g_{D1}, g_{D2}], as well as production conditions in public offices in the regions, determine the budget volume. Outputs, budgets, utilities of bureaucrats, and votes in both regions may be derived.

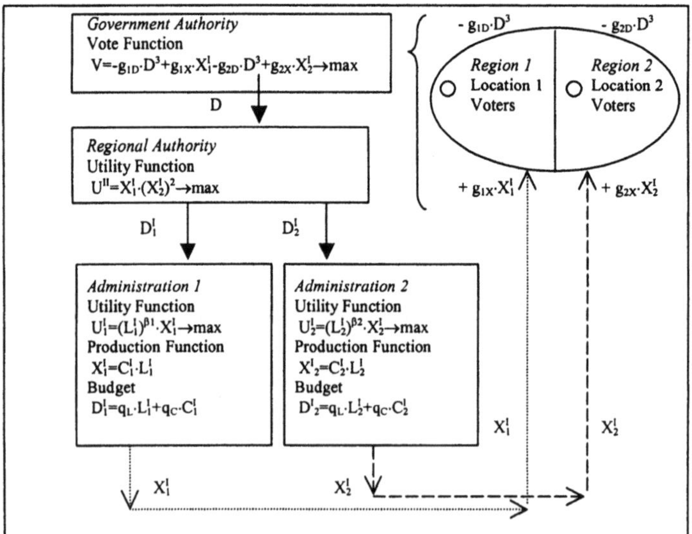

The maximisation of the votes for both regions yields the optimal budget

$$D_{Opt} = \frac{2 \cdot g_{1X} \cdot \frac{1+\beta_1}{(2+\beta_1)^2} + 8 \cdot g_{2X} \cdot \frac{1+\beta_2}{(2+\beta_2)^2}}{27 \cdot q_L \cdot q_C \cdot (g_{1D} + g_{2D})}$$

considering the equations above as maximisation of utilities at rank I and rank II administrations.

Fig. 5.3 Regional political competition within an administrative sector [for the definition of the terms, see Notation]

The model outlined in Fig. 5.3 shows a solution for sector specific political administrative competition. If the sensitivity of voters to the financial burden increases, the optimal budget shrinks. The same happens in the case of increases in factor prices. Greater sensitivity of voters to output increases the budget, whereas more preferences for labour inputs by managers in low-rank administrations allow the budget to shrink. An extension of the total budget increases the output in region, especially as it benefits from preferences of the second rank administration for that region. Budget cuts are more at the expense of region 1. The model can be easily extended, considering different factor prices in regions, negotiations between low-rank administrations, the production of services by higher rank public offices, the separation of the utility functions of top managers in rank III public office and of politicians in parliament, other regions, and non-financial competition. Institutional regional competition is very much influenced by the evaluations of low rank administrative management, by regional preferences, and evaluations of the middle rank management and by the voting behaviour in the regions. Outcome of regional competition is also affected by the assessment of the total budget.

5.5 Political Regional Competition within a Federation

An Extended Administrative Model with Equal Reaction of Voters in Federal and State Elections

The model outlined in Fig. 5.3 may be developed further in several directions. In particular, we will leave the simple two-region world by introducing a federal level [rank III administration]. A rank II administration denotes the state [provincial] level. The model structure is given in Fig. 5.4. Note that there is an administrative branch which is not only under the power of one jurisdiction alone. Consequently, it is necessary to introduce some more general production functions, utility functions of low rank administrations and vote functions at the state level. This is described mathematically in Fig. 5.6 where we distinguish between three cases. Case (a) refers to identical voting functions at the top and middle levels (i.e. II and III) while cases (b) and (c) involve different voting functions. Case (b) differs from case (c) by the assumption that there is the same ruling party at levels II and III. Fig. 5.5 characterises a function that relates the output of the model to alternative sizes of an individual public office.

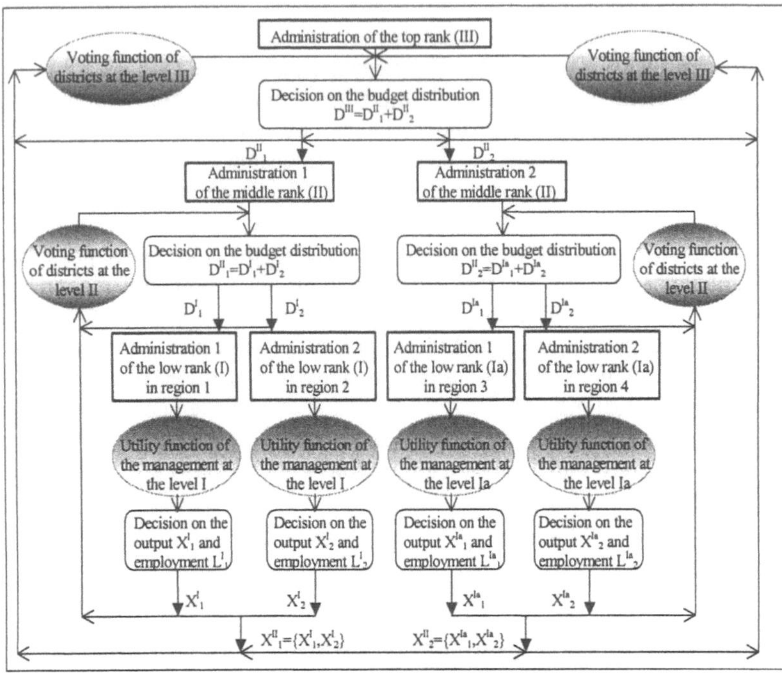

Fig. 5.4 Administrative sector model with equal reaction of voters in federal and state elections [for the definition of the terms, see Notation]

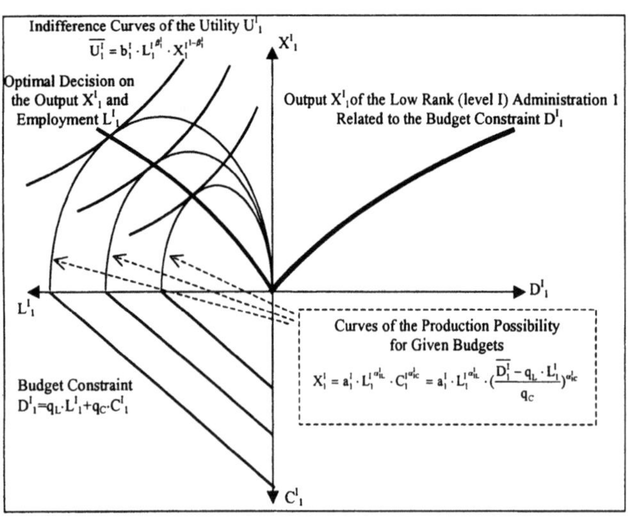

Fig. 5.5 Low rank administration [for the definition of the terms, see Notation]

Low rank administration, for example: low rank administration 1 at level I Production function (2) $X_1^I = a_1^I \cdot (L_1^I)^{\alpha_{1L}^I} \cdot (C_1^I)^{\alpha_{1C}^I}$, where $\alpha_{1L}^I > 0$, $\alpha_{1C}^I > 0$, $\alpha_{1L}^I + \alpha_{1C}^I = \alpha_1^I < 1$ Budget constraint (3) $D_1^I = q_L \cdot L_1^I + q_C \cdot C_1^I \geq 0$ Utility function of the management (4) $U_1^I = b_1^I \cdot (L_1^I)^{\beta_1^I} \cdot (X_1^I)^{1-\beta_1^I}$ where $0 < \beta_1^I < 1$
Middle rank administration (state/province/county), for example: administration 1 at the middle level II Budget constraint (5) $D_1^{II} = D_1^I + D_2^I$ where $1 = \frac{dD_1^I}{dD_1^{II}} + \frac{dD_2^I}{dD_1^{II}}$ Voting function in districts of the low rank administrations at level I (6) $V_1^I = g_{1X}^I \cdot X_1^I - g_{1D}^I \cdot D_1^I \geq 0$, $V_2^I = g_{2X}^I \cdot X_2^I - g_{2D}^I \cdot D_2^I \geq 0$, where $g_{1X}^I > 0$, $g_{1D}^I > 0$, $g_{2X}^I > 0$, $g_{2D}^I > 0$ Voting function in the entire district of the middle rank administration (7) $V_1^{II} = V_1^I + V_2^I \geq 0$
Top rank administration (Federation) Budget constraint (8) $D^{III} = D_1^{II} + D_2^{II}$
Case a): Identical voting functions at the top and middle levels (9) $V^{III} = V_1^{II} + V_2^{II}$
Case b): Different voting functions at the top and middle administrative levels, same ruling party at the top level III and the middle level II Voting functions in the districts of the middle rank administrations 1 and 2 (10) $V_1^{III} = g_{1X}^{II} \cdot X_1^{II} - g_{1D1}^{II} \cdot D_1^{II} - g_{1D2}^{II} \cdot D_2^{II} \geq 0$ $V_2^{III} = g_{2X}^{II} \cdot X_2^{II} - g_{2D2}^{II} \cdot D_2^{II} - g_{2D1}^{II} \cdot D_1^{II} \geq 0$ where $X_1^{II} = X_1^I + X_2^I$, $X_2^{II} = X_1^{Ia} + X_2^{Ia}$, $D_1^{II} = D_1^I + D_2^I$, $D_2^{II} = D_1^{Ia} + D_2^{Ia}$ $g_{1X}^{II} > 0$, $g_{2X}^{II} > 0$, $g_{1D2}^{II} > g_{1D1}^{II} > 0$, $g_{2D1}^{II} > g_{2D2}^{II} > 0$ Votes out of the entire district of the top rank administration (11) $V^{III} = V_1^{III} + V_2^{III}$
Case c): Different voting functions at the top and middle administrative levels, different ruling parties at the top and the middle levels Voting functions in the districts of the middle rank administrations 1 and 2 (12) $V_1^{III} = g_{1X}^{II} \cdot X_1^{II} - g_{1D1}^{II} \cdot D_1^{II} - g_{1D2}^{II} \cdot D_2^{II} \geq 0$ $V_2^{III} = g_{2X}^{II} \cdot X_2^{II} - g_{2D2}^{II} \cdot D_2^{II} - g_{2D1}^{II} \cdot D_1^{II} \geq 0$ Votes out of the entire district of the top rank administration (13) $V^{III} = V_1^{III} + (V_{2N}^{III} - V_2^{III})$, where V_2^{III} denotes votes for the opposition P2 at the top level III, $V_{2N}^{III} - V_2^{III}$ are the votes for the ruling party P1 at the top level III and $V_{2N}^{III} - V_2^{III} \geq 0$.

Fig. 5.6 Description of administrations [for the definition of terms, see Notation]

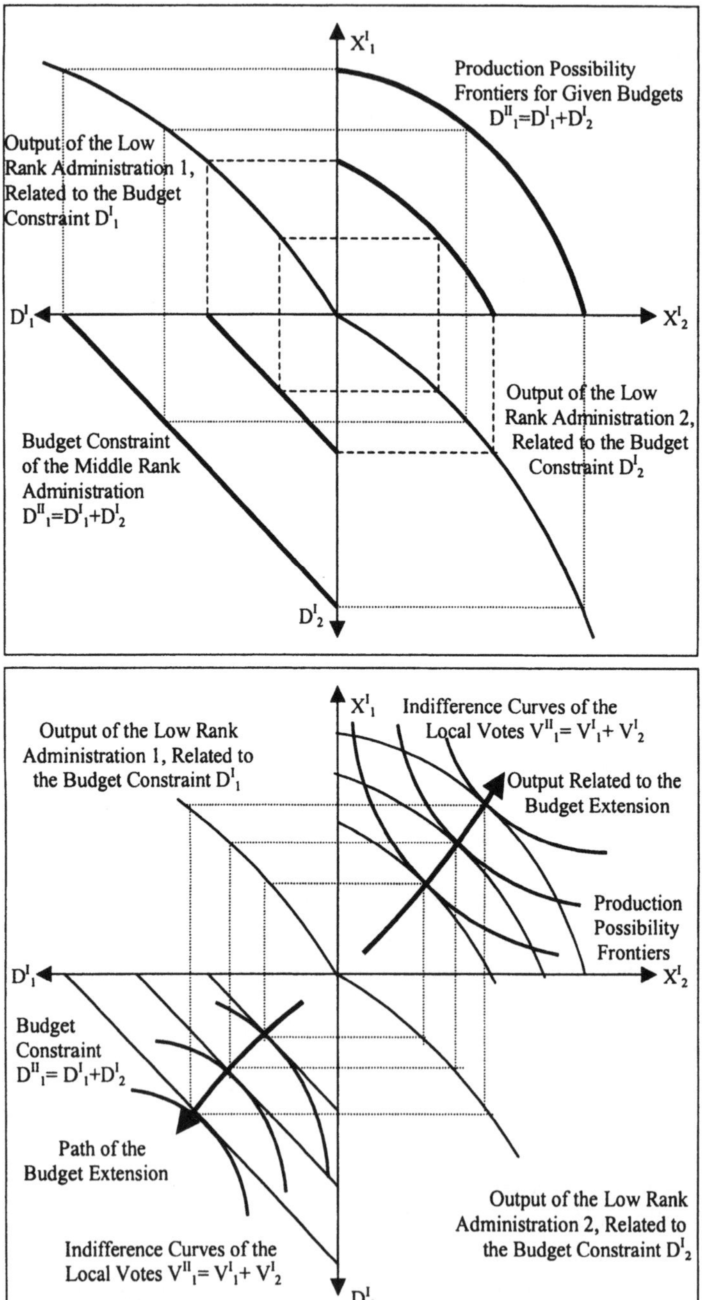

Fig. 5.7 Budget distribution through middle rank administration [for the definition of terms, see Notation]

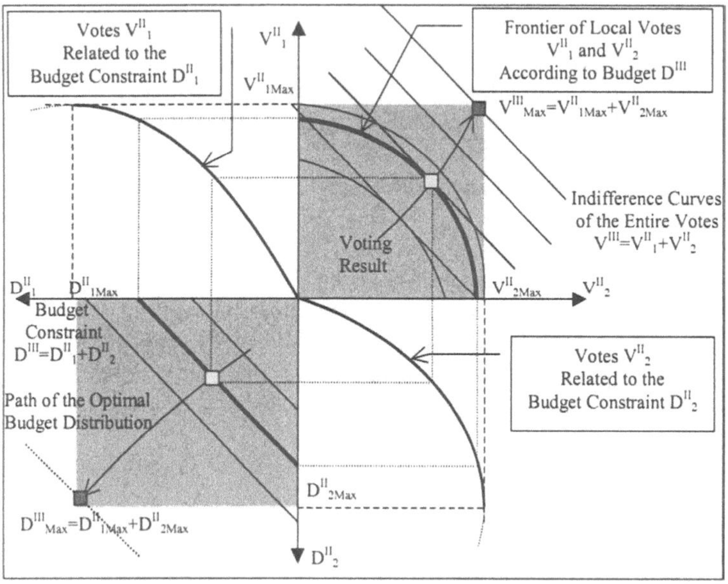

Fig. 5.8 Top rank administration: Case a) [for the definition of terms, see Notation]

At level II [state level] a jurisdiction distributes a budget amongst the low level administrations in such a way as to maximise voters' support. Two jurisdictions [states] are shown, each of them controlling two low level administrations. For a given budget allocated to each middle rank jurisdiction, a production possibility frontier concerning the outputs of the related low rank administrations, is derived (as shown in Fig. 5.5). Vote functions concerning the middle rank administration, stress the dependency of voters judgements on outputs and budget sizes. A vote function is derived, showing the results of voting related to the outputs of the low rank administrations for a given budget. A set of indifference curves demonstrates equal votes. In Fig. 5.7 the optimal budget distribution is fixed.

At level III [federal level] the budget distribution among the second level jurisdictions and the total budget, become determined. The distribution of budget at level II establishes a relationship between their budget and the output produced by the low level administrations, which maximises votes in the territories of middle rank jurisdictions. If the behaviour of voters in federal elections is the same as in the regional elections, the federation distributes the budget in such a way that federal elections and local elections are won (for more details see Equation (17), outlined in Fig. 5.11). It also fixes the total budget in such a way that the total budget amounts to a size which maximises the votes at the middle rank level (for more details see Equation (15), outlined in Fig. 5.11). This is demonstrated by Fig. 5.8.

There is a tendency to create a considerably large total budget. Regional competition for budgets leads to high budgets and high activities in competing regions. As seen above, variations in factor prices, production functions, preferences of voters etc., influence the results of regional competition.

An Extended Administrative Model with Different Reactions of Voters in Federal and State Votes

A similar model may be constructed where the same political, federal and administrative structure is assumed. But we consider different vote functions, as in Fig. 5.7 we assumed that the voters in the regions react to political regional competition. They dislike the budget increase for administrative purposes. The budget of the competing regions are evaluated negatively. This is expressed in the vote functions presented in Fig. 5.9.

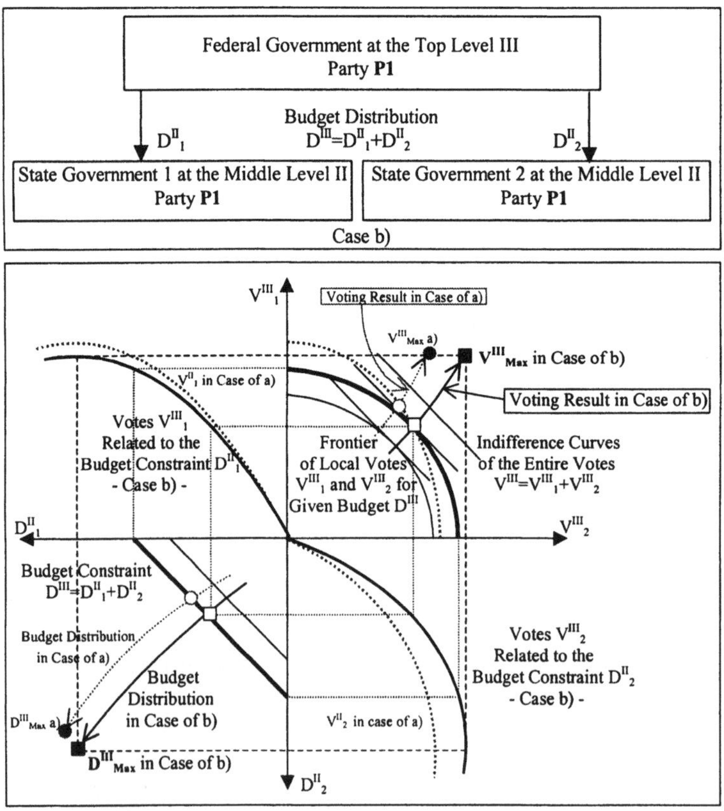

Fig. 5.9 Different parties in power in federal and state government [for the definition of terms, see Notation]

The total budget which results turns out to be less than in the case where less envy and less consciousness of regional competition for votes is prevalent (see Equation (16) in Fig 5.11). As shown in Fig. 5.9, the vote functions undergo a partly downward shift leading to a smaller total budget. In such cases, regional competition may lead to smaller budgets and less preferable results. Moreover, one can elaborate on the impacts of parameter changes within this framework of political regional competition.

An Extended Administrative Model with Different Parties in Governments

A further variant of the approach involves different vote functions at the middle rank jurisdiction level, which differ from those at the federal level. We also assume that not all jurisdictions show the same party in government. It is supposed that in one of the middle rank jurisdictions the opposition is in government and voters of this region switch to the opposition if improvements occur, e.g. with respect to output (see Fig. 5.10).

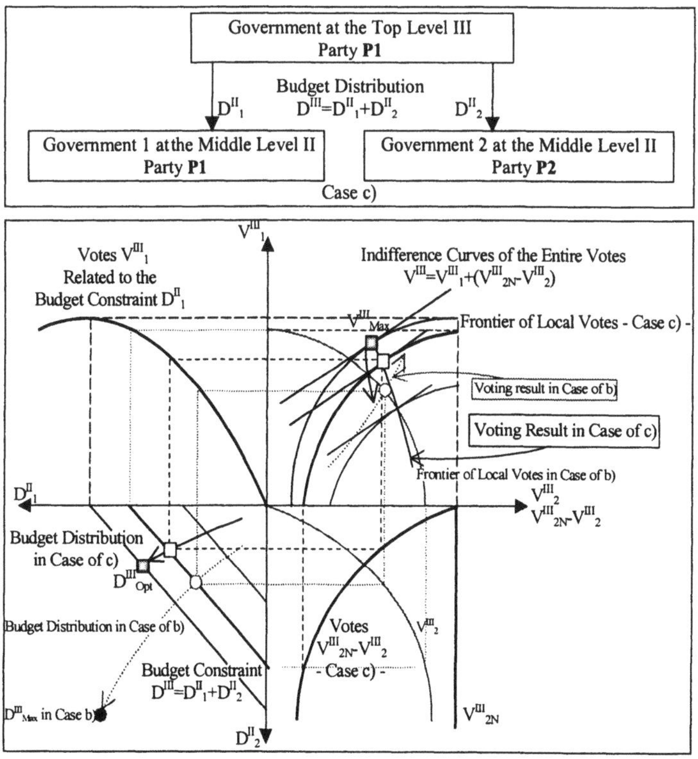

Fig. 5.10 Top rank administration: Case c) [for the definition of terms, see Notation]

Top Rank Administration (Federation)

Case a):

(14)
$$D^{III}=\left(\frac{\alpha^{I}_{1}\cdot g^{I}_{1X}\cdot x^{I}_{1}}{g^{I}_{1D}}\right)^{\frac{1}{1-\alpha^{I}_{1}}}+\left(\frac{\alpha^{I}_{2}\cdot g^{I}_{2X}\cdot x^{I}_{2}}{g^{I}_{2D}}\right)^{\frac{1}{1-\alpha^{I}_{2}}}+\left(\frac{\alpha^{Ia}_{1}\cdot g^{Ia}_{1X}\cdot x^{Ia}_{1}}{g^{Ia}_{1D}}\right)^{\frac{1}{1-\alpha^{Ia}_{1}}}+\left(\frac{\alpha^{Ia}_{2}\cdot g^{Ia}_{2X}\cdot x^{Ia}_{2}}{g^{Ia}_{2D}}\right)^{\frac{1}{1-\alpha^{Ia}_{2}}}$$

Case b):

(15)
$$D^{III}_{Max}=(g^{II}_{1D2}+g^{II}_{2D1})\cdot$$
$$\cdot\frac{\left(\frac{g^{I}_{1D}-g^{I}_{2D}}{1-\varepsilon^{I}_{D1,2}}\cdot\frac{g^{II}_{1X}}{g^{I}_{1X}}-g^{II}_{1D1}+g^{II}_{1D2}\right)\cdot\delta^{II}_{1,1}+\left(\frac{g^{I}_{2D}-g^{I}_{1D}}{1-\varepsilon^{I}_{D2,1}}\cdot\frac{g^{II}_{1X}}{g^{I}_{2X}}-g^{II}_{1D1}+g^{II}_{1D2}\right)\cdot\delta^{II}_{1,2}+\left(\frac{g^{Ia}_{1D}-g^{Ia}_{2D}}{1-\varepsilon^{Ia}_{D1,2}}\cdot\frac{g^{II}_{2X}}{g^{Ia}_{1X}}-g^{II}_{2D2}+g^{II}_{2D1}\right)\cdot\delta^{II}_{2,1}+\left(\frac{g^{Ia}_{2D}-g^{Ia}_{1D}}{1-\varepsilon^{Ia}_{D2,1}}\cdot\frac{g^{II}_{2X}}{g^{Ia}_{2X}}-g^{II}_{2D2}+g^{II}_{2D1}\right)\cdot\delta^{II}_{2,2}}{\left[\left(\frac{g^{I}_{1D}-g^{I}_{2D}}{1-\varepsilon^{I}_{D1,2}}\cdot\frac{g^{II}_{1X}}{g^{I}_{1X}}-g^{II}_{1D1}+g^{II}_{1D2}\right)\cdot\delta^{II}_{1,1}+\left(\frac{g^{I}_{2D}-g^{I}_{1D}}{1-\varepsilon^{I}_{D2,1}}\cdot\frac{g^{II}_{1X}}{g^{I}_{2X}}-g^{II}_{1D1}+g^{II}_{1D2}\right)\right]\cdot\left[\left(\frac{g^{Ia}_{1D}-g^{Ia}_{2D}}{1-\varepsilon^{Ia}_{D1,2}}\cdot\frac{g^{II}_{2X}}{g^{Ia}_{1X}}-g^{II}_{2D2}+g^{II}_{2D1}\right)\cdot\delta^{II}_{2,1}+\left(\frac{g^{Ia}_{2D}-g^{Ia}_{1D}}{1-\varepsilon^{Ia}_{D2,1}}\cdot\frac{g^{II}_{2X}}{g^{Ia}_{2X}}-g^{II}_{2D2}+g^{II}_{2D1}\right)\cdot\delta^{II}_{2,2}\right]}$$

Case c):

(16)
$$D^{III}_{Opt}=(g^{II}_{1D2}-g^{II}_{2D1})\cdot$$
$$\frac{\left(\frac{g^{I}_{1D}-g^{I}_{2D}}{1-\varepsilon^{I}_{D1,2}}\cdot\frac{g^{II}_{1X}}{g^{I}_{1X}}-g^{II}_{1D1}+g^{II}_{1D2}\right)\cdot\delta^{II}_{1,1}+\left(\frac{g^{I}_{2D}-g^{I}_{1D}}{1-\varepsilon^{I}_{D2,1}}\cdot\frac{g^{II}_{1X}}{g^{I}_{2X}}-g^{II}_{1D1}+g^{II}_{1D2}\right)\cdot\delta^{II}_{1,2}+\left(g^{II}_{2D2}-g^{II}_{2D1}-\frac{g^{Ia}_{1D}-g^{Ia}_{2D}}{1-\varepsilon^{Ia}_{D1,2}}\cdot\frac{g^{II}_{2X}}{g^{Ia}_{1X}}\right)\cdot\delta^{II}_{2,1}-\left(g^{II}_{2D2}-g^{II}_{2D1}-\frac{g^{Ia}_{2D}-g^{Ia}_{1D}}{1-\varepsilon^{Ia}_{D2,1}}\cdot\frac{g^{II}_{2X}}{g^{Ia}_{2X}}\right)\cdot\delta^{II}_{2,2}}{\left[\left(\frac{g^{I}_{1D}-g^{I}_{2D}}{1-\varepsilon^{I}_{D1,2}}\cdot\frac{g^{II}_{1X}}{g^{I}_{1X}}-g^{II}_{1D1}+g^{II}_{1D2}\right)\cdot\delta^{II}_{1,1}+\left(\frac{g^{I}_{2D}-g^{I}_{1D}}{1-\varepsilon^{I}_{D2,1}}\cdot\frac{g^{II}_{1X}}{g^{I}_{2X}}-g^{II}_{1D1}+g^{II}_{1D2}\right)\cdot\delta^{II}_{1,2}\right]\cdot\left[\left(g^{II}_{2D2}-g^{II}_{2D1}-\frac{g^{Ia}_{1D}-g^{Ia}_{2D}}{1-\varepsilon^{Ia}_{D1,2}}\cdot\frac{g^{II}_{2X}}{g^{Ia}_{1X}}\right)\cdot\delta^{II}_{2,1}-\left(g^{II}_{2D2}-g^{II}_{2D1}-\frac{g^{Ia}_{2D}-g^{Ia}_{1D}}{1-\varepsilon^{Ia}_{D2,1}}\cdot\frac{g^{II}_{2X}}{g^{Ia}_{2X}}\right)\cdot\delta^{II}_{2,2}\right]}$$

Middle Rank Administration (State)

Case a):

(17)
$$D^{II}_{1\,Max}=\left(\frac{\alpha^{I}_{1}\cdot g^{I}_{1X}\cdot x^{I}_{1}}{g^{I}_{1D}}\right)^{\frac{1}{1-\alpha^{I}_{1}}}+\left(\frac{\alpha^{I}_{2}\cdot g^{I}_{2X}\cdot x^{I}_{2}}{g^{I}_{2D}}\right)^{\frac{1}{1-\alpha^{I}_{2}}},\quad D^{II}_{2\,Max}=\left(\frac{\alpha^{Ia}_{1}\cdot g^{Ia}_{1X}\cdot x^{Ia}_{1}}{g^{Ia}_{1D}}\right)^{\frac{1}{1-\alpha^{Ia}_{1}}}+\left(\frac{\alpha^{Ia}_{2}\cdot g^{Ia}_{2X}\cdot x^{Ia}_{2}}{g^{Ia}_{2D}}\right)^{\frac{1}{1-\alpha^{Ia}_{2}}}$$

Case b):

(18)
$$D^{II}_{1Max}=$$
$$\frac{(g^{II}_{1D2}+g^{II}_{2D1})\cdot\left[\left(\frac{g^{Ia}_{1D}-g^{Ia}_{2D}}{1-\varepsilon^{Ia}_{D1,2}}\cdot\frac{g^{II}_{2X}}{g^{Ia}_{1X}}-g^{II}_{2D2}+g^{II}_{2D1}\right)\cdot\delta^{II}_{2,1}+\left(\frac{g^{Ia}_{2D}-g^{Ia}_{1D}}{1-\varepsilon^{Ia}_{D2,1}}\cdot\frac{g^{II}_{2X}}{g^{Ia}_{2X}}-g^{II}_{2D2}+g^{II}_{2D1}\right)\cdot\delta^{II}_{2,2}\right]}{\left[\left(\frac{g^{I}_{1D}-g^{I}_{2D}}{1-\varepsilon^{I}_{D1,2}}\cdot\frac{g^{II}_{1X}}{g^{I}_{1X}}-g^{II}_{1D1}+g^{II}_{1D2}\right)\cdot\delta^{II}_{1,1}+\left(\frac{g^{I}_{2D}-g^{I}_{1D}}{1-\varepsilon^{I}_{D2,1}}\cdot\frac{g^{II}_{1X}}{g^{I}_{2X}}-g^{II}_{1D1}+g^{II}_{1D2}\right)\cdot\delta^{II}_{1,2}\right]\cdot\left[\left(\frac{g^{Ia}_{1D}-g^{Ia}_{2D}}{1-\varepsilon^{Ia}_{D1,2}}\cdot\frac{g^{II}_{2X}}{g^{Ia}_{1X}}-g^{II}_{2D2}+g^{II}_{2D1}\right)\cdot\delta^{II}_{2,1}+\left(\frac{g^{Ia}_{2D}-g^{Ia}_{1D}}{1-\varepsilon^{Ia}_{D2,1}}\cdot\frac{g^{II}_{2X}}{g^{Ia}_{2X}}-g^{II}_{2D2}+g^{II}_{2D1}\right)\cdot\delta^{II}_{2,2}\right]}$$

$$D^{II}_{2Max}=$$
$$\frac{(g^{II}_{1D2}+g^{II}_{2D1})\cdot\left[\left(\frac{g^{I}_{1D}-g^{I}_{2D}}{1-\varepsilon^{I}_{D1,2}}\cdot\frac{g^{II}_{1X}}{g^{I}_{1X}}-g^{II}_{1D1}+g^{II}_{1D2}\right)\cdot\delta^{II}_{1,1}+\left(\frac{g^{I}_{2D}-g^{I}_{1D}}{1-\varepsilon^{I}_{D2,1}}\cdot\frac{g^{II}_{1X}}{g^{I}_{2X}}-g^{II}_{1D1}+g^{II}_{1D2}\right)\cdot\delta^{II}_{1,2}\right]}{\left[\left(\frac{g^{I}_{1D}-g^{I}_{2D}}{1-\varepsilon^{I}_{D1,2}}\cdot\frac{g^{II}_{1X}}{g^{I}_{1X}}-g^{II}_{1D1}+g^{II}_{1D2}\right)\cdot\delta^{II}_{1,1}+\left(\frac{g^{I}_{2D}-g^{I}_{1D}}{1-\varepsilon^{I}_{D2,1}}\cdot\frac{g^{II}_{1X}}{g^{I}_{2X}}-g^{II}_{1D1}+g^{II}_{1D2}\right)\cdot\delta^{II}_{1,2}\right]\cdot\left[\left(\frac{g^{Ia}_{1D}-g^{Ia}_{2D}}{1-\varepsilon^{Ia}_{D1,2}}\cdot\frac{g^{II}_{2X}}{g^{Ia}_{1X}}-g^{II}_{2D2}+g^{II}_{2D1}\right)\cdot\delta^{II}_{2,1}+\left(\frac{g^{Ia}_{2D}-g^{Ia}_{1D}}{1-\varepsilon^{Ia}_{D2,1}}\cdot\frac{g^{II}_{2X}}{g^{Ia}_{2X}}-g^{II}_{2D2}+g^{II}_{2D1}\right)\cdot\delta^{II}_{2,2}\right]}$$

Case c):

(19)
$$D^{II}_{1Opt}=$$
$$\frac{(g^{II}_{1D2}-g^{II}_{2D1})\cdot\left[\left(g^{II}_{2D2}-g^{II}_{2D1}-\frac{g^{Ia}_{1D}-g^{Ia}_{2D}}{1-\varepsilon^{Ia}_{D1,2}}\cdot\frac{g^{II}_{2X}}{g^{Ia}_{1X}}\right)\cdot\delta^{II}_{2,1}-\left(g^{II}_{2D2}-g^{II}_{2D1}-\frac{g^{Ia}_{2D}-g^{Ia}_{1D}}{1-\varepsilon^{Ia}_{D2,1}}\cdot\frac{g^{II}_{2X}}{g^{Ia}_{2X}}\right)\cdot\delta^{II}_{2,2}\right]}{\left[\left(\frac{g^{I}_{1D}-g^{I}_{2D}}{1-\varepsilon^{I}_{D1,2}}\cdot\frac{g^{II}_{1X}}{g^{I}_{1X}}-g^{II}_{1D1}+g^{II}_{1D2}\right)\cdot\delta^{II}_{1,1}+\left(\frac{g^{I}_{2D}-g^{I}_{1D}}{1-\varepsilon^{I}_{D2,1}}\cdot\frac{g^{II}_{1X}}{g^{I}_{2X}}-g^{II}_{1D1}+g^{II}_{1D2}\right)\cdot\delta^{II}_{1,2}\right]\cdot\left[\left(g^{II}_{2D2}-g^{II}_{2D1}-\frac{g^{Ia}_{1D}-g^{Ia}_{2D}}{1-\varepsilon^{Ia}_{D1,2}}\cdot\frac{g^{II}_{2X}}{g^{Ia}_{1X}}\right)\cdot\delta^{II}_{2,1}-\left(g^{II}_{2D2}-g^{II}_{2D1}-\frac{g^{Ia}_{2D}-g^{Ia}_{1D}}{1-\varepsilon^{Ia}_{D2,1}}\cdot\frac{g^{II}_{2X}}{g^{Ia}_{2X}}\right)\cdot\delta^{II}_{2,2}\right]}$$

$$D^{II}_{2Opt}=$$
$$\frac{(g^{II}_{1D2}-g^{II}_{2D1})\cdot\left[\left(\frac{g^{I}_{1D}-g^{I}_{2D}}{1-\varepsilon^{I}_{D1,2}}\cdot\frac{g^{II}_{1X}}{g^{I}_{1X}}-g^{II}_{1D1}+g^{II}_{1D2}\right)\cdot\delta^{II}_{1,1}+\left(\frac{g^{I}_{2D}-g^{I}_{1D}}{1-\varepsilon^{I}_{D2,1}}\cdot\frac{g^{II}_{1X}}{g^{I}_{2X}}-g^{II}_{1D1}+g^{II}_{1D2}\right)\cdot\delta^{II}_{1,2}\right]}{\left[\left(\frac{g^{I}_{1D}-g^{I}_{2D}}{1-\varepsilon^{I}_{D1,2}}\cdot\frac{g^{II}_{1X}}{g^{I}_{1X}}-g^{II}_{1D1}+g^{II}_{1D2}\right)\cdot\delta^{II}_{1,1}+\left(\frac{g^{I}_{2D}-g^{I}_{1D}}{1-\varepsilon^{I}_{D2,1}}\cdot\frac{g^{II}_{1X}}{g^{I}_{2X}}-g^{II}_{1D1}+g^{II}_{1D2}\right)\cdot\delta^{II}_{1,2}\right]\cdot\left[\left(g^{II}_{2D2}-g^{II}_{2D1}-\frac{g^{Ia}_{1D}-g^{Ia}_{2D}}{1-\varepsilon^{Ia}_{D1,2}}\cdot\frac{g^{II}_{2X}}{g^{Ia}_{1X}}\right)\cdot\delta^{II}_{2,1}-\left(g^{II}_{2D2}-g^{II}_{2D1}-\frac{g^{Ia}_{2D}-g^{Ia}_{1D}}{1-\varepsilon^{Ia}_{D2,1}}\cdot\frac{g^{II}_{2X}}{g^{Ia}_{2X}}\right)\cdot\delta^{II}_{2,2}\right]}$$

Low Rank Administration

Case a):

(20)
$$D^{I}_{1\,Max}=\left(\frac{\alpha^{I}_{1}\cdot g^{I}_{1X}\cdot x^{I}_{1}}{g^{I}_{1D}}\right)^{\frac{1}{1-\alpha^{I}_{1}}},\quad D^{I}_{2\,Max}=\left(\frac{\alpha^{I}_{2}\cdot g^{I}_{2X}\cdot x^{I}_{2}}{g^{I}_{2D}}\right)^{\frac{1}{1-\alpha^{I}_{2}}},\quad D^{Ia}_{1Max}=\left(\frac{\alpha^{Ia}_{1}\cdot g^{Ia}_{1X}\cdot x^{Ia}_{1}}{g^{Ia}_{1D}}\right)^{\frac{1}{1-\alpha^{Ia}_{1}}},\quad D^{Ia}_{2Max}=\left(\frac{\alpha^{Ia}_{2}\cdot g^{Ia}_{2X}\cdot x^{Ia}_{2}}{g^{Ia}_{2D}}\right)^{\frac{1}{1-\alpha^{Ia}_{2}}}$$

ctd.

Case b)

(21) $D^{I}_{1Max}=$

$$\frac{\frac{\delta^{II}_{1,1}}{\varepsilon^{I}_{D1,2}{}'}\cdot(1-\alpha^{I}_{1})\cdot(\varepsilon^{I}_{D1,2}-1)\cdot(g^{II}_{1D2}+g^{II}_{2D1})\cdot\left[\left(\frac{g^{Ia}_{1D}-g^{Ia}_{2D}}{1-\varepsilon^{Ia}_{D1,2}}\cdot\frac{g^{II}_{2X}}{g^{Ia}_{1X}}-g^{II}_{2D2}+g^{II}_{2D1}\right)\cdot\delta^{II}_{2,1}+\left(\frac{g^{Ia}_{2D}-g^{Ia}_{1D}}{1-\varepsilon^{Ia}_{D2,1}}\cdot\frac{g^{II}_{2X}}{g^{Ia}_{2X}}-g^{II}_{2D2}+g^{II}_{2D1}\right)\cdot\delta^{II}_{2,2}\right]}{\left[\left(\frac{g^{I}_{1D}-g^{I}_{2D}}{1-\varepsilon^{I}_{D1,2}}\cdot\frac{g^{II}_{1X}}{g^{I}_{1X}}-g^{II}_{1D1}+g^{II}_{1D2}\right)\cdot\delta^{II}_{1,1}+\left(\frac{g^{I}_{2D}-g^{I}_{1D}}{1-\varepsilon^{I}_{D2,1}}\cdot\frac{g^{II}_{1X}}{g^{I}_{2X}}-g^{II}_{1D1}+g^{II}_{1D2}\right)\cdot\delta^{II}_{1,2}\right]\cdot\left[\left(\frac{g^{Ia}_{1D}-g^{Ia}_{2D}}{1-\varepsilon^{Ia}_{D1,2}}\cdot\frac{g^{II}_{2X}}{g^{Ia}_{1X}}-g^{II}_{2D2}+g^{II}_{2D1}\right)\cdot\delta^{II}_{2,1}+\left(\frac{g^{Ia}_{2D}-g^{Ia}_{1D}}{1-\varepsilon^{Ia}_{D2,1}}\cdot\frac{g^{II}_{2X}}{g^{Ia}_{2X}}-g^{II}_{2D2}+g^{II}_{2D1}\right)\cdot\delta^{II}_{2,2}\right]}$$

$D^{I}_{2Max}=$

$$\frac{\frac{\delta^{II}_{1,2}}{\varepsilon^{I}_{D2,1}{}'}\cdot(1-\alpha^{I}_{2})\cdot(\varepsilon^{I}_{D2,1}-1)\cdot(g^{II}_{1D2}+g^{II}_{2D1})\cdot\left[\left(\frac{g^{Ia}_{1D}-g^{Ia}_{2D}}{1-\varepsilon^{Ia}_{D1,2}}\cdot\frac{g^{II}_{2X}}{g^{Ia}_{1X}}-g^{II}_{2D2}+g^{II}_{2D1}\right)\cdot\delta^{II}_{2,1}+\left(\frac{g^{Ia}_{2D}-g^{Ia}_{1D}}{1-\varepsilon^{Ia}_{D2,1}}\cdot\frac{g^{II}_{2X}}{g^{Ia}_{2X}}-g^{II}_{2D2}+g^{II}_{2D1}\right)\cdot\delta^{II}_{2,2}\right]}{\left[\left(\frac{g^{I}_{1D}-g^{I}_{2D}}{1-\varepsilon^{I}_{D1,2}}\cdot\frac{g^{II}_{1X}}{g^{I}_{1X}}-g^{II}_{1D1}+g^{II}_{1D2}\right)\cdot\delta^{II}_{1,1}+\left(\frac{g^{I}_{2D}-g^{I}_{1D}}{1-\varepsilon^{I}_{D2,1}}\cdot\frac{g^{II}_{1X}}{g^{I}_{2X}}-g^{II}_{1D1}+g^{II}_{1D2}\right)\cdot\delta^{II}_{1,2}\right]\cdot\left[\left(\frac{g^{Ia}_{1D}-g^{Ia}_{2D}}{1-\varepsilon^{Ia}_{D1,2}}\cdot\frac{g^{II}_{2X}}{g^{Ia}_{1X}}-g^{II}_{2D2}+g^{II}_{2D1}\right)\cdot\delta^{II}_{2,1}+\left(\frac{g^{Ia}_{2D}-g^{Ia}_{1D}}{1-\varepsilon^{Ia}_{D2,1}}\cdot\frac{g^{II}_{2X}}{g^{Ia}_{2X}}-g^{II}_{2D2}+g^{II}_{2D1}\right)\cdot\delta^{II}_{2,2}\right]}$$

$D^{Ia}_{1Max}=$

$$\frac{\frac{\delta^{II}_{2,1}}{\varepsilon^{Ia}_{D1,2}{}'}\cdot(1-\alpha^{Ia}_{1})\cdot(\varepsilon^{Ia}_{D1,2}-1)\cdot(g^{II}_{1D2}+g^{II}_{2D1})\cdot\left[\left(\frac{g^{I}_{1D}-g^{I}_{2D}}{1-\varepsilon^{I}_{D1,2}}\cdot\frac{g^{II}_{1X}}{g^{I}_{1X}}-g^{II}_{1D1}+g^{II}_{1D2}\right)\cdot\delta^{II}_{1,1}+\left(\frac{g^{I}_{2D}-g^{I}_{1D}}{1-\varepsilon^{I}_{D2,1}}\cdot\frac{g^{II}_{1X}}{g^{I}_{2X}}-g^{II}_{1D1}+g^{II}_{1D2}\right)\cdot\delta^{II}_{1,2}\right]}{\left[\left(\frac{g^{I}_{1D}-g^{I}_{2D}}{1-\varepsilon^{I}_{D1,2}}\cdot\frac{g^{II}_{1X}}{g^{I}_{1X}}-g^{II}_{1D1}+g^{II}_{1D2}\right)\cdot\delta^{II}_{1,1}+\left(\frac{g^{I}_{2D}-g^{I}_{1D}}{1-\varepsilon^{I}_{D2,1}}\cdot\frac{g^{II}_{1X}}{g^{I}_{2X}}-g^{II}_{1D1}+g^{II}_{1D2}\right)\cdot\delta^{II}_{1,2}\right]\cdot\left[\left(\frac{g^{Ia}_{1D}-g^{Ia}_{2D}}{1-\varepsilon^{Ia}_{D1,2}}\cdot\frac{g^{II}_{2X}}{g^{Ia}_{1X}}-g^{II}_{2D2}+g^{II}_{2D1}\right)\cdot\delta^{II}_{2,1}+\left(\frac{g^{Ia}_{2D}-g^{Ia}_{1D}}{1-\varepsilon^{Ia}_{D2,1}}\cdot\frac{g^{II}_{2X}}{g^{Ia}_{2X}}-g^{II}_{2D2}+g^{II}_{2D1}\right)\cdot\delta^{II}_{2,2}\right]}$$

$D^{Ia}_{2Max}=$

$$\frac{\frac{\delta^{II}_{2,2}}{\varepsilon^{Ia}_{D2,1}{}'}\cdot(1-\alpha^{Ia}_{2})\cdot(\varepsilon^{Ia}_{D2,1}-1)\cdot(g^{II}_{1D2}+g^{II}_{2D1})\cdot\left[\left(\frac{g^{I}_{1D}-g^{I}_{2D}}{1-\varepsilon^{I}_{D1,2}}\cdot\frac{g^{II}_{1X}}{g^{I}_{1X}}-g^{II}_{1D1}+g^{II}_{1D2}\right)\cdot\delta^{II}_{1,1}+\left(\frac{g^{I}_{2D}-g^{I}_{1D}}{1-\varepsilon^{I}_{D2,1}}\cdot\frac{g^{II}_{1X}}{g^{I}_{2X}}-g^{II}_{1D1}+g^{II}_{1D2}\right)\cdot\delta^{II}_{1,2}\right]}{\left[\left(\frac{g^{I}_{1D}-g^{I}_{2D}}{1-\varepsilon^{I}_{D1,2}}\cdot\frac{g^{II}_{1X}}{g^{I}_{1X}}-g^{II}_{1D1}+g^{II}_{1D2}\right)\cdot\delta^{II}_{1,1}+\left(\frac{g^{I}_{2D}-g^{I}_{1D}}{1-\varepsilon^{I}_{D2,1}}\cdot\frac{g^{II}_{1X}}{g^{I}_{2X}}-g^{II}_{1D1}+g^{II}_{1D2}\right)\cdot\delta^{II}_{1,2}\right]\cdot\left[\left(\frac{g^{Ia}_{1D}-g^{Ia}_{2D}}{1-\varepsilon^{Ia}_{D1,2}}\cdot\frac{g^{II}_{2X}}{g^{Ia}_{1X}}-g^{II}_{2D2}+g^{II}_{2D1}\right)\cdot\delta^{II}_{2,1}+\left(\frac{g^{Ia}_{2D}-g^{Ia}_{1D}}{1-\varepsilon^{Ia}_{D2,1}}\cdot\frac{g^{II}_{2X}}{g^{Ia}_{2X}}-g^{II}_{2D2}+g^{II}_{2D1}\right)\cdot\delta^{II}_{2,2}\right]}$$

Case c):

(22) $D^{I}_{1Opt}=$

$$\frac{\frac{\delta^{II}_{1,1}}{\varepsilon^{I}_{D1,2}{}'}\cdot(1-\alpha^{I}_{1})\cdot(\varepsilon^{I}_{D1,2}-1)\cdot(g^{II}_{1D2}+g^{II}_{2D1})\cdot\left[\left(g^{II}_{2D2}-g^{II}_{2D1}-\frac{g^{Ia}_{1D}-g^{Ia}_{2D}}{1-\varepsilon^{Ia}_{D1,2}}\cdot\frac{g^{II}_{2X}}{g^{Ia}_{1X}}\right)\cdot\delta^{II}_{2,1}-\left(g^{II}_{2D2}-g^{II}_{2D1}-\frac{g^{Ia}_{2D}-g^{Ia}_{1D}}{1-\varepsilon^{Ia}_{D2,1}}\cdot\frac{g^{II}_{2X}}{g^{Ia}_{2X}}\right)\cdot\delta^{II}_{2,2}\right]}{\left[\left(\frac{g^{I}_{1D}-g^{I}_{2D}}{1-\varepsilon^{I}_{D1,2}}\cdot\frac{g^{II}_{1X}}{g^{I}_{1X}}-g^{II}_{1D1}+g^{II}_{1D2}\right)\cdot\delta^{II}_{1,1}+\left(\frac{g^{I}_{2D}-g^{I}_{1D}}{1-\varepsilon^{I}_{D2,1}}\cdot\frac{g^{II}_{1X}}{g^{I}_{2X}}-g^{II}_{1D1}+g^{II}_{1D2}\right)\cdot\delta^{II}_{1,2}\right]\cdot\left[\left(g^{II}_{2D2}-g^{II}_{2D1}-\frac{g^{Ia}_{1D}-g^{Ia}_{2D}}{1-\varepsilon^{Ia}_{D1,2}}\cdot\frac{g^{II}_{2X}}{g^{Ia}_{1X}}\right)\cdot\delta^{II}_{2,1}-\left(g^{II}_{2D2}-g^{II}_{2D1}-\frac{g^{Ia}_{2D}-g^{Ia}_{1D}}{1-\varepsilon^{Ia}_{D2,1}}\cdot\frac{g^{II}_{2X}}{g^{Ia}_{2X}}\right)\cdot\delta^{II}_{2,2}\right]}$$

$D^{I}_{2Opt}=$

$$\frac{\frac{\delta^{II}_{1,2}}{\varepsilon^{I}_{D2,1}{}'}\cdot(1-\alpha^{I}_{2})\cdot(\varepsilon^{I}_{D2,1}-1)\cdot(g^{II}_{1D2}-g^{II}_{2D1})\cdot\left[\left(g^{II}_{2D2}-g^{II}_{2D1}-\frac{g^{Ia}_{1D}-g^{Ia}_{2D}}{1-\varepsilon^{Ia}_{D1,2}}\cdot\frac{g^{II}_{2X}}{g^{Ia}_{1X}}\right)\cdot\delta^{II}_{2,1}-\left(g^{II}_{2D2}-g^{II}_{2D1}-\frac{g^{Ia}_{2D}-g^{Ia}_{1D}}{1-\varepsilon^{Ia}_{D2,1}}\cdot\frac{g^{II}_{2X}}{g^{Ia}_{2X}}\right)\cdot\delta^{II}_{2,2}\right]}{\left[\left(\frac{g^{I}_{1D}-g^{I}_{2D}}{1-\varepsilon^{I}_{D1,2}}\cdot\frac{g^{II}_{1X}}{g^{I}_{1X}}-g^{II}_{1D1}+g^{II}_{1D2}\right)\cdot\delta^{II}_{1,1}+\left(\frac{g^{I}_{2D}-g^{I}_{1D}}{1-\varepsilon^{I}_{D2,1}}\cdot\frac{g^{II}_{1X}}{g^{I}_{2X}}-g^{II}_{1D1}+g^{II}_{1D2}\right)\cdot\delta^{II}_{1,2}\right]\cdot\left[\left(g^{II}_{2D2}-g^{II}_{2D1}-\frac{g^{Ia}_{1D}-g^{Ia}_{2D}}{1-\varepsilon^{Ia}_{D1,2}}\cdot\frac{g^{II}_{2X}}{g^{Ia}_{1X}}\right)\cdot\delta^{II}_{2,1}-\left(g^{II}_{2D2}-g^{II}_{2D1}-\frac{g^{Ia}_{2D}-g^{Ia}_{1D}}{1-\varepsilon^{Ia}_{D2,1}}\cdot\frac{g^{II}_{2X}}{g^{Ia}_{2X}}\right)\cdot\delta^{II}_{2,2}\right]}$$

$D^{Ia}_{1Opt}=$

$$\frac{\frac{\delta^{II}_{2,1}}{\varepsilon^{Ia}_{D1,2}{}'}\cdot(1-\alpha^{Ia}_{1})\cdot(\varepsilon^{Ia}_{D1,2}-1)\cdot(g^{II}_{1D2}-g^{II}_{2D1})\cdot\left[\left(\frac{g^{I}_{1D}-g^{I}_{2D}}{1-\varepsilon^{I}_{D1,2}}\cdot\frac{g^{II}_{1X}}{g^{I}_{1X}}-g^{II}_{1D1}+g^{II}_{1D2}\right)\cdot\delta^{II}_{1,1}+\left(\frac{g^{I}_{2D}-g^{I}_{1D}}{1-\varepsilon^{I}_{D2,1}}\cdot\frac{g^{II}_{1X}}{g^{I}_{2X}}-g^{II}_{1D1}+g^{II}_{1D2}\right)\cdot\delta^{II}_{1,2}\right]}{\left[\left(\frac{g^{I}_{1D}-g^{I}_{2D}}{1-\varepsilon^{I}_{D1,2}}\cdot\frac{g^{II}_{1X}}{g^{I}_{1X}}-g^{II}_{1D1}+g^{II}_{1D2}\right)\cdot\delta^{II}_{1,1}+\left(\frac{g^{I}_{2D}-g^{I}_{1D}}{1-\varepsilon^{I}_{D2,1}}\cdot\frac{g^{II}_{1X}}{g^{I}_{2X}}-g^{II}_{1D1}+g^{II}_{1D2}\right)\cdot\delta^{II}_{1,2}\right]\cdot\left[\left(g^{II}_{2D2}-g^{II}_{2D1}-\frac{g^{Ia}_{1D}-g^{Ia}_{2D}}{1-\varepsilon^{Ia}_{D1,2}}\cdot\frac{g^{II}_{2X}}{g^{Ia}_{1X}}\right)\cdot\delta^{II}_{2,1}-\left(g^{II}_{2D2}-g^{II}_{2D1}-\frac{g^{Ia}_{2D}-g^{Ia}_{1D}}{1-\varepsilon^{Ia}_{D2,1}}\cdot\frac{g^{II}_{2X}}{g^{Ia}_{2X}}\right)\cdot\delta^{II}_{2,2}\right]}$$

$D^{Ia}_{2Opt}=$

$$\frac{\frac{\delta^{II}_{2,2}}{\varepsilon^{Ia}_{D2,1}{}'}\cdot(1-\alpha^{Ia}_{2})\cdot(\varepsilon^{Ia}_{D2,1}-1)\cdot(g^{II}_{1D2}-g^{II}_{2D1})\cdot\left[\left(\frac{g^{I}_{1D}-g^{I}_{2D}}{1-\varepsilon^{I}_{D1,2}}\cdot\frac{g^{II}_{1X}}{g^{I}_{1X}}-g^{II}_{1D1}+g^{II}_{1D2}\right)\cdot\delta^{II}_{1,1}+\left(\frac{g^{I}_{2D}-g^{I}_{1D}}{1-\varepsilon^{I}_{D2,1}}\cdot\frac{g^{II}_{1X}}{g^{I}_{2X}}-g^{II}_{1D1}+g^{II}_{1D2}\right)\cdot\delta^{II}_{1,2}\right]}{\left[\left(\frac{g^{I}_{1D}-g^{I}_{2D}}{1-\varepsilon^{I}_{D1,2}}\cdot\frac{g^{II}_{1X}}{g^{I}_{1X}}-g^{II}_{1D1}+g^{II}_{1D2}\right)\cdot\delta^{II}_{1,1}+\left(\frac{g^{I}_{2D}-g^{I}_{1D}}{1-\varepsilon^{I}_{D2,1}}\cdot\frac{g^{II}_{1X}}{g^{I}_{2X}}-g^{II}_{1D1}+g^{II}_{1D2}\right)\cdot\delta^{II}_{1,2}\right]\cdot\left[\left(g^{II}_{2D2}-g^{II}_{2D1}-\frac{g^{Ia}_{1D}-g^{Ia}_{2D}}{1-\varepsilon^{Ia}_{D1,2}}\cdot\frac{g^{II}_{2X}}{g^{Ia}_{1X}}\right)\cdot\delta^{II}_{2,1}-\left(g^{II}_{2D2}-g^{II}_{2D1}-\frac{g^{Ia}_{2D}-g^{Ia}_{1D}}{1-\varepsilon^{Ia}_{D2,1}}\cdot\frac{g^{II}_{2X}}{g^{Ia}_{2X}}\right)\cdot\delta^{II}_{2,2}\right]}$$

Fig. 5.11 Solutions for budgets of administrations [for the definition of terms, see Notation]

The method of deriving solutions is the same as in the other cases. At the top level, the total budget is no longer equal to vote maximising state budgets. A set of optimal budget distributions for alternative budget sizes is elaborated (see Fig. 5.10 and Fig. 5.11 for more details). Out of this set, the one which maximises votes for the federal level is determined (see Equation (17) in Fig. 5.11). The solution is graphically shown in Fig. 5.10 and mathematically given by Fig. 5.11. The total budget fixed is smaller than that in the case of equal governments in all jurisdictions. Regional competition does lead to smaller overall administrative outputs than it does in the other cases. The jurisdiction which has the same party in government as at the federal level, receives a relatively higher budget and therefore benefits from a preferential position in regional competition. Again, many changes can be introduced to investigate the impacts of and on political regional competition. The approaches may be extended to more detailed revenues of jurisdictions, other means of administrative coordination, etc. Tax revenues and public debts are considered in an approach to macroeconomic regional fiscal competition.

5.6 Political Fiscal Regional Competition and Political Competition of Municipalities and Public Firms

Already in the 1970s van Rompuy (1977) developed a model for horizontal regional *competition by fiscal policy.* More recently Friedrich (1987) adapted this model to cases of regional stagnation and vertical regional competition. He extended the model for a federation with horizontal competition among states [regions] and vertical competition between the federation and the states [regions], maximising income of the individual regions[3]. Regional governments and federal governments try to control the economic situation by fiscal policies. They can apply three different strategies: a Haavelmo balanced budget policy, a deficit spending policy and a laisser faire policy. According to the different or common fiscal policies applied within the region and by the federation, the results in terms of the income of region 1 and of region 2 are different[4]. Two pay off matrices can be elaborated and the regions can be interpreted as players representing region 1 and region 2.

From this perspective the regions may be viewed as adopting the income maximising strategies. If the situation on the capital markets in both regions is different in so far that for region 1 severe crowding out effects occur, while region 2 benefits from a rich capital market, then the regions adopt dominant strategies. Region 1 will play Haavelmo and region 2 deficit spending, thus experiencing different fiscal policies. The federation is introduced as a player of fiscal policy, too. It comes out that both regions benefit from the federal fiscal policy, but to different extents. Consequently, the choice of fiscal strategies in horizontal fiscal policy competition is not influenced. This result differs from the findings by the administration hierarchy model presented above. In the above model income maximising governments are assumed to belong to the same political party. The federal government shows no regional preferences which would lead to combining the choice of a fiscal strategy with sanctions or support if individual regional governments played specific strategies. If regional government 1 is formed by a party which is in favour of low prices, while regional government 2 by a party related to voters who appreciate high incomes, the choice of strategies may vary. Region 1 also turns to a Haavelmo strategy. Berndt (1979) has introduced into such a framework the assumption that voters in one state consider not only their experience with the government of their own state, but also experiences with governments in other states too.

In some countries the basic parameter of action and instrument of municipal settlement competition is zoning. The territory is planned and those real estates dedicated to industrial or sometimes administrative use are going to be sold for settlement to private companies or the public offices of other jurisdictions. After evaluating the results of settlement policy, the planners may rezone the land use, according to new expectations (Lindemann, 1999). At first, we may assume that settlement competition takes place within given land use plans which allow for settlements at the locations in question.

This can be modelled in various ways. One is to formulate a game where a public or private investor looks for appropriate *locations in municipalities*, e.g. two cities. The investor may have an order of preference with respect to specified possible locations in the two cities. Their strategy is to offer locations to attract investors. According to the offers of the two competing municipalities, the preferences of the investor and the impact of the location, result in the choice of the investor by two pay off matrices, one for each municipality. By the application of game theoretic solution concepts, the best strategies of the municipalities and the resulting location are determined. The pay offs can be formulated in political terms such as votes related to the settlement (Friedrich, 1976, pp. 287). The settlement may lead to a new composition of the voting body, or the settlement may change voting behaviour. If in the two cities, the political party in government expects favourable gains in votes from the settlement, both cities are in severe conflict for the same settlement. The usual results of settlement games can be expected, e.g. the maximisation of jobs in the cities. If the city governments belong to different political parties, groups etc. however, a very different location can be chosen compared to the first case. One city might prefer to avoid a settlement, although it may be forced by the state to join the game. Therefore, the investor is not able to find a highly preferred location. If the political outcome of settlement turns out for both municipalities to be politically negative, e.g. in the case of the settlement of military units or atomic power stations, and if municipalities are obliged to join the game, both try to offer locations which are best preferred by the investor, in order to avoid highly negative political pay offs.

This approach is open to the possibility of considering many instruments of regional competition and many kinds of political conflicts (Friedrich, 1976).

Distribution of i Possible Locations in Town 1 ($S_{1,i}$) and j Possible Locations in Town 2 ($S_{2,j}$)

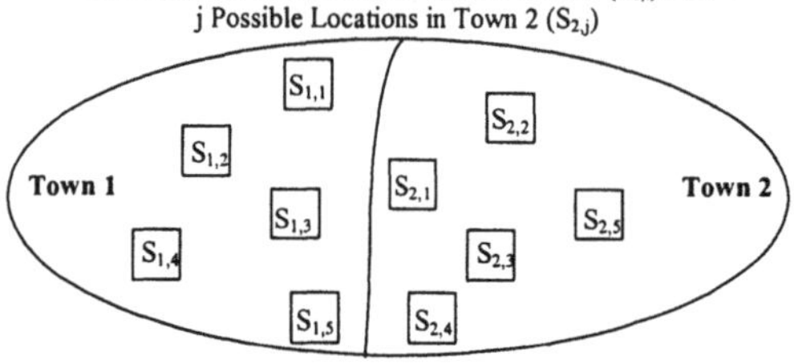

Preferences of the Investor: $S_{1,1}>S_{2,1}>S_{1,5}>S_{2,2}>S_{2,5}>S_{2,4}>S_{1,2}>S_{1,3}>S_{1,4}>S_{2,3}$

Result of Location Choice with Regard to Alternative Location Offers of Town 1 and Town 2

		Town 2					
		$S_{2,1}$	$S_{2,2}$	$S_{2,3}$	$S_{2,4}$	$S_{2,5}$	
Town 1	$S_{1,1}$	$S_{1,1}$	$S_{1,1}$	$S_{1,1}$	$S_{1,1}$	$S_{1,1}$	$S_{1,1}$
	$S_{1,2}$	$S_{2,1}$	$S_{2,2}$	$S_{1,2}$	$S_{2,4}$	$S_{2,5}$	$S_{1,2}$
	$S_{1,3}$	$S_{2,1}$	$S_{2,2}$	$S_{1,3}$	$S_{2,4}$	$S_{2,5}$	$S_{1,3}$
	$S_{1,4}$	$S_{2,1}$	$S_{2,2}$	$S_{1,4}$	$S_{2,4}$	$S_{2,5}$	$S_{1,4}$
	$S_{1,5}$	$S_{2,1}$	$S_{1,5}$	$S_{1,5}$	$S_{1,5}$	$S_{1,5}$	$S_{1,5}$

Pay Off in Votes

		Town 2				
		$S_{2,1}$	$S_{2,2}$	$S_{2,3}$	$S_{2,4}$	$S_{2,5}$
Town 1	$S_{1,1}$	300, 200	300, 200	300, 200	300, 200	300, 200
	$S_{1,2}$	150, 350	100, 400	320, 180	10, 490	5, 495
	$S_{1,3}$	150, 350	100, 400	380, 120	10, 490	5, 495
	$S_{1,4}$	150, 350	100, 400	480, 20	10, 490	5, 495
	$S_{1,5}$	150, 350	490, 10	490, 10	490, 10	490, 10

Distribution of Changes in Votes Resulting from Different Locations

Town 1		Town 2	
$S_{1,1}$	300, 200	$S_{2,1}$	150, 350
$S_{1,2}$	320, 180	$S_{2,2}$	100, 400
$S_{1,3}$	380, 120	$S_{2,3}$	95, 405
$S_{1,4}$	480, 20	$S_{2,4}$	10, 490
$S_{1,5}$	490, 10	$S_{2,5}$	5, 495

Source: Friedrich (1976, pp. 288-290)

Fig. 5.12 Example of a political location game [for the definition of terms, see Notation]

Another approach to political business promotion competition covers more than the results of a one-shot game. We assume, in an already zoned municipal region, that a municipal *promotion agency negotiating with a private firm* looks for a *location*. The municipality is able to sell locations to the investor and to offer various financial incentives which lead to different purchase prices for the investor. The city government maximises political support by voters. The private investor is interested in high profits. Firstly, we suppose negotiations to be carried on between one city promotion institution and one investor. Secondly, it is assumed that the private investor negotiates with a competing municipality as well. A model recently developed by Friedrich and Feng (2000) aims to find a solution to this bargaining situation.

Voters in a municipality react in favour of the governing party if the number of jobs, the investments and the revenues from the sale along with an improvement of environment due to subsidies are positive. Moreover, the business concept of the investor related to the output of the firm is evaluated positively by voters if output rises. They also react positively to local government policy if the budget increases with the price of the location sold to the investor. This acceptance shrinks if the costs of sale and payments of subsidies increase. The reaction of voters is expressed in a *goal function* of local government.

As the *investor maximises profit*, he aims to minimise costs. The change in votes [utility of town] and the profit of the firm depend on the production volume and the price of real estate[5]. A Nash-solution of negotiations between town and investors is determined. The *change in votes* depends on the parameters. Higher sensitivity of voters regarding output, jobs and investment increases the political outcome. Higher taxes, again, show a tendency for losing voters' support to diminish.

The *price of real estate* is reduced when voters increase their sensitivity towards output, more jobs, higher investments and subsidies, and when prices rise and demand for the investors' products shrinks. Greater elasticity of the settlement firm increases the real estate price. If minimum utility or alternative investment possibilities improve, the real estate price decreases. In the case where the alternative of the local government to attract voters increases, the real estate price also increases. These conditions and competitive actions of regional competition are considered in the model.

In the case of competing local governments the investor can negotiate with the *competing municipality as well*, thus achieving a solution there (Friedrich and Feng, 2000). The investor may choose the most favourable solution, opening a new round of negotiations, using the best result achieved with one municipality as a minimum condition in negotiation with the other. This will result in improved conditions until one municipality has reached the minimum requirement and does not confirm to bargain. Such competition processes are tough if both localities have the same type of political government, or if voters react in both municipalities in the same way. If voter movement in one municipality does not benefit government to a large extent, the investor has fewer possibilities of pressurising the competing municipality by increasing his minimum requirements from one round to the next. It may happen that the market form is reduced to a bilateral monopoly for political reasons. In the political regional competition to attract investors, one finds bilateral monopolies, supply restricted monopsonies, demand restricted monopolies, supply restricted oligopsonies and demand restricted oligopolies (Friedrich, 1977; Lindemann, 1999). Bilateral oligopolies.are very rare.

Lindemann (1999) has suggested, the exhibition of regional competition in land use planning. This can easily be transferred to political competition. As mentioned, settlement competition through negotiations with private and public investors is embedded in the framework of territorial and land use planning. Therefore, zoning is the most important parameter of action in general municipal competition[6]. There are three types of competition models:

- a model of competition by land use planning,
- the models of individual sales of real estate for industrial settlement, and
- a model of adaptation of land use plans.

Models of the first und third type can be extended to political regional competition as well.

It is assumed that neighbouring municipalities or municipalities of the same rank compete in the zoning phase. They usually meet each other under oligopolistic horizontal market conditions. For the sake of simplicity, a duopoly is assumed. We suppose that voters react positively to the output of the private sector and the public sector. For the sake of simplicity here also an additive vote function is introduced. The votes in both communities are maximised under considerations of mutual dependencies[7].

Another extension of the Lindemann model refers to an urban land use adaptation model. Here again a vote function, as mentioned above, is introduced. The basic structure of the model is not changed (Friedrich and Lindemann, 2000). The municipal decision makers compare the achieved average bargaining real estate price, with their expected average revenue from sales. The future expectations adapt to this result according to a reaction function. The community does not react if the difference between the expected and the achieved average revenue is zero. The revenue function of one or both municipalities change and new solutions can be derived.

The approach gives room to determine solution changes in horizontal regional competition, if parameters such as wages, taxes, production conditions, and voters' sensitivities change. If in one community voters react more positively to public production, the land for industrial purposes will be reduced. Lower financial means can increase voters well being. It also determines the effect of the outcome of vertical negotiations between municipality and settlement firms in various market forms, on zoning and the consequences of horizontal competition in zoning, through the revision of the urban territorial plans. Thus, impacts of many factors influence business promotion negotiations on zoning. It is possible to model interrelations among horizontal competition between communities, horizontal competition among firms for locations, horizontal competition of location suppliers and vertical negotiations. The approach serves to detect the effects of changing parameters, fixed in negotiations with higher ranking public offices or jurisdictions, such as tax rates or subsidies from the European Union.

There are few models that *study regional competitive activities* of public enterprises[8]. In particular, the horizontal and vertical competition explicitly related to competing regions is insufficiently studied. Politically motivated decisions were analysed in relation to the achievement of economic goals important for politicians (Thiemeyer, 1970, 1975; Eichhorn, 1978; Oettle, 1976; Friedrich, 1969, 1970; Bös, 1991; Rees, 1993). See also the models of politically oriented price, or investment policy of public enterprises, by Peltzman (1971), Bös (1989) and Ziemes (1992). Moreover, we would like to draw the reader's attention to the analysis provided by Friedrich and Feng (2000) in which regional competition is considered in association with political goals.

Management of public enterprises detects actions in horizontal competition related to sales, procurement, production and some financial activities themselves. The public owner determines vertically the tasks of the enterprise in regional competition. Usually the owner [principal], e.g. a city, and the firm's management do not fix individual goals of competition, but have to specify the financial scope for the firm's activities. With respect to substantial investment decisions financial constraints are estimated. This financial target depends on the competitive position of a municipality and its goals of competition. Thus, the financial constraint is assumed to depend on the competition situation and it also depends on the political goals of the public owner.

In the model of Friedrich and Feng (2000) it is assumed that a municipality and a municipal enterprise negotiate the fixing of the financial scope as well as the output to be produced. The *utility of a municipality* depends on the attraction of voters through the achievement of a high output of the municipal firm and on its financial contribution to the municipal budget. The utility of management of the public firm depends on output and employment. Both community and public enterprises seek to realise, each for itself, a *minimum utility level*[9]. Using this model the bargaining *solution by Nash*, the stipulated financial contribution, the fixed output, and the product prise can be seen (Friedrich and Feng, 2000, p. 195).

If the financial contribution is of greater importance in strengthening the competitive position of a region, a high municipal tendency for competition will lead to a high value of financial municipal returns in the case of firms where budget orientation is important and voters accept high budget contributions. With rising competition this value increases. But, if under strongly competitive circumstances the municipal owner employs the public enterprise directly for competitive purposes, the municipality is interested in high output. A highly competitive situation corresponds to a low value of financial returns. Therefore, higher competition leads to a smaller value of financial contributions to the municipality. A new Nash solution results. A tendency to a lower *financial contribution to the municipality* leads to an increasing output. The lower price for the services of the public firm prevails. According to the Nash solutions, a public owner's demand for a higher financial contribution, results in lower output for a higher price.

5.7 Summary and Outlook

The outcome of regional competition is very much influenced by regional competition among public institutions and their actions. The role of public offices and public enterprises of jurisdictions in regional competition can be determined by referring to public regional competitors, the kind of regional competition, basic horizontal and vertical competition relations, instruments of regional competition, market forms of regional competition, goals of regional competition, and results of regional competition.

For competing public offices and firms a rather complex net of regional horizontal and vertical competition emerges. This net cannot be modelled easily. Essential features of regional competition are not considered in existing models of the competition of firms, nor are models of fiscal federalism discussed in public finance. They do not comprise the microeconomic and macroeconomic determinants of regional competition. Moreover, they neglect the relevant kind of regional competition, the public offices and public firms as decision makers, the market forms and the political and administrative goals of the public regional competitors. Identification of implications of competitive measures on the outcome of regional competition in one model is rather difficult. All the various kinds of regional competition cannot be modelled in one approach. Therefore, we have concentrated on the political competition of public offices only.

A starting point offers the coordination theory of public offices by the application of simple game theory. A model of an administrative branch with three levels of public offices and decision-makers following political goals or utility functions of management within a financial coordination process allows the determination of regional competition results with respect to outputs, factor inputs, votes, satisfaction of management and voters in regions, and budgets in regions, as well as the overall budget of the administrative branch.

A second model group serves to clarify competition among jurisdictions following political goals in a federation. A federal state, provinces [states], and low level administrations exist. Political competition prevails. Again relevant variables of regional competition can be determined. Two versions of the model are offered. In one model, voters show the same voting behaviour in federal elections and state elections and the same party is in power in all governments considered. In another model, the evaluation of voters with respect to state and federal elections are different and different parties are in power in different jurisdictions. The models show the results on regional competition. In some situations the outputs, budgets, etc. turn out to be smaller than in the first case. Changes in favour of regions are traced where the same political party is in power as at the federal level.

Another group of models refer to fiscal competition within a federation where the jurisdictions try to achieve regional income maximisation or vote maximisation. Federal fiscal policies benefit the regions differently. As long as these policies are not related to fiscal policy measures of the states their competitive behaviour is not influenced.

Political models of the horizontal settlement competition between firms describe game-like situations of business promotion policy, and refer to some typical outcomes differing according to which party is in power in different states and municipalities and the political evaluation functions of the politicians. One version refers to bilateral vertical monopoly when selling real estates. Business promotion models which refer to zoning competition, competition by sales of real estates, and revisions of zoning are translated into political competition. They apply to bilateral monopolies as well as to supply-restricted monopsonies.

Finally, a model of negotiations between the public proprietor and the management of public enterprises on determining actions in regional competition is presented. The public owner is going to achieve political goals. Depending on the goals and the role of public enterprise in regional competition, different outputs and prices as well as budget contributions of the public firm result. The findings are helpful in order to formulate more relevant concepts of regional competition, regional growth models and indexes of regional competition.

Endnotes

1 Other goals are to maximise an ideology function, staff politically relevant posts, gain majorities in decision making bodies, logrolling among parties, provide services and goods to benefit supporters (Renault and Winden, 1991), achieve a higher rank for a decision making body, gain power in territorial and functional reforms, enlarge competencies and available resources of the institutions (Friedrich, 1976).

2 There are very few exceptions (Friedrich, 1976; Berndt, 1979; Friedrich, 1987; Breton, 1996; Gerken, 1999, Christiansen, 2000) where *political factors* in regional competition are taken into account. Some roots of considering political aspects stem from a federalism theory (Frey, 1977; Kirsch, 1977; Thöni, 1986; Epple and Zelenitz, 1981; Wildasin, 1986; Oates, 1991; Goudriaan and de Groot, 1991; Renault and Winden, 1991; Arnold, 1992; Postlep, 1993; Frey, 1997; Zimmermann, 1999; Schwager, 2000).

3 For mathematical derivation of the results the reader should turn to Friedrich (1987). Moreover, goods movements, capital movements, and labour movements among regions, price level and production functions, were introduced. Federation and two states exist at a regional level. A macroeconomic description of the economy of the two regions, with a goods market, a capital market, a rudimentary real labour and production sector and the public sectors of the regions exists.

4 These outcomes show the multipliers caused by combinations of policies of region 2 and region 1. They can be interpreted as pay offs in a game between the regions where the regional governments attempt to maximise income in their own regions.

5 Therefore, indifference curves of the town and the investor can be traced. The actors negotiate on both the real estate curve and further sales conditions. The points of tangency show a pareto-optimal sequence combining real estate price and output

(Friedrich and Feng, 2000, pp. 27). If the real estate price increases, the investor reduces his output. To each pareto-optimal combination a combination of the actors utilities can be found to correspond. By application of the *solution concept of Nash*, under conditions of minimum utilities to be achieved by town and investor, the solution may be found (Friedrich and Feng, 1995; Friedrich and Feng, 1998, Friedrich and Feng, 2000). Thus, the real estate price, the vote change, the profit and output, input and subsidy is determined.

6 Therefore, on the municipal level, three phase municipal competition with respect to settlement competition takes place. At first there is the preparation of land use planning under consideration by the competing municipalities. Finally, a phase of competition and negotiation of individual settlements follows, as characterised above. A phase of adaptation of land use planning on the basis of experiences gained through the results of individual settlement negotiations, has recently begun.

7 For private output, a production function of the Cobb-Douglas type exists which connects private sector output with labour input and the proportion of industrial area (zone) planned. The public output is produced according to a non-substitutive production function, where output depends on labour productivity, real wages, real revenues from sales of real estate and from real tax revenues. Expected sales revenues depend on its own land price levels and on the estimated price level for industrial real estate prices [average revenue per m^2] in the competing municipality. Tax revenues depend on real income gained in the community and on the value of industrial zones, as well as tax rates. It is also assumed that final private production is reduced by real tax payments. These dependencies are substituted in such a way that a dependency among the average sales price expectation of one community depends on that of the other municipality. Demand functions will be elaborated for each of the municipalities. Under the behavioural assumption by Launhardt and Hotelling, an oligopolistic solution of the duopoly is found which determines the land use plan of both municipalities. The procedures needed to derive a Launhardt solution are found in Lindemann (1999) and in Friedrich and Lindemann (2000). Under different behaviour of the communities, a Stackelberg solution as well a collusion solution may result. The number of real estates ready for sales, can be fixed.

8 Literature on public enterprises considers their role as instruments of economic policy (Thiemeyer, 1970). They are interpreted as economic units for the improvement of social welfare, or as factors for the development of an economy according to Ritschl (Ritschl, 1931; Friedrich, 1987; Hirsch, 1992).

9 The *situation of the public firm* shows production, demand, cost, and finance functions. An area for possible negotiation solutions referring to financial contributions, the volume of production, and the respective utilities of negotiators is determined.

Notation

a_i^I	Technical parameter of the production function of low rank administration I at location i (i = 1, 2)
$a_j^{I_a}$	Technical parameter of the production function of low rank administration I_a at location j (j = 1, 2)

b_i^I	Level parameter of the utility evaluation of low rank administration I at location i (i = 1, 2)
$b_j^{I_a}$	Level parameter of the utility evaluation of low rank administration I_a at location j (j = 1, 2)
C	Production factor capital
C_i^I	Capital stock of low rank administration I at location i (i = 1, 2)
$C_j^{I_a}$	Capital stock of low rank administration I_a at location j (j = 1, 2)
D	Budget of an administration
D_i^I	Budget of low rank administration I at location i (i = 1, 2)
D_{iMax}^I	Budget of low rank administration I at location i (i = 1, 2) resulted from maximisation of votes
D_{iOpt}^I	Budget of low rank administration I at location i (i = 1, 2) on the path of the optimal budget extension
$D_j^{I_a}$	Budget of low rank administration I_a at location j (j = 1, 2)
$D_{jMax}^{I_a}$	Budget of low rank administration I_a at location j (j = 1, 2) resulted from maximisation of votes
$D_{jOpt}^{I_a}$	Budget of low rank administration I_a at location j (j = 1, 2) on the path of the optimal budget extension
D_k^{II}	Budget of middle rank administration II in region k (k = 1, 2)
D_{kMax}^{II}	Budget of middle rank administration II in region k (k = 1, 2) resulted from maximisation of votes
D_{kOpt}^{II}	Budget of middle rank administration II in region k (k = 1, 2) on the path of the optimal budget extension
D^{III}	Budget of top rank administration III
D_{Max}^{III}	Budget of top rank administration III resulted from maximisation of votes
D_{Opt}^{III}	Budget of top rank administration III on the path of the optimal budget extension
g_{iX}^I	Parameter of the voting function with regard to output of low rank administration I at location i (i = 1, 2)
$g_{jX}^{I_a}$	Parameter of the voting function with regard to output of low rank administration I_a at location j (j = 1, 2)
g_{iD}^I	Parameter of the voting function with regard to budget of low rank administration I at location i (i = 1, 2)
$g_{jD}^{I_a}$	Parameter of the voting function with regard to budget of low rank administration I_a at location j (j = 1, 2)
g_{jX}^{II}	Parameter of the voting function with regard to outputs of middle rank administration II in region j (j = 1, 2)
g_{jDk}^{II}	Parameter of the voting function with regard to budget of middle rank administration II in region j (k = 1, 2)
L	Production factor labour
L_i^I	Labour allocation of low rank administration I at location i (i = 1, 2)
$L_j^{I_a}$	Labour allocation of low rank administration I_a at location j (j = 1, 2)
P1	Ruling party at middle level II (state) and the top level III (federation)
P2	Ruling party at middle level III (state), but opposition at the top level III (federation)
q_C, q_L	Price of production factors
$S_{i,j}$	Possible location j (j = 1, ..., 5) in town i (i = 1, 2)
U	Utility function of the management of an administration
U_i^I	Utility function of the management of low rank administration I at location i (i = 1, 2)
$U_j^{I_a}$	Utility function of the management of low rank administration I_a at location j (j = 1, 2)
V	Voting function in the district of an administration
V_i^I	Voting function in district i (i = 1, 2) of low rank administration I
$V_j^{I_a}$	Voting function in district j (j = 1, 2) of low rank administration I_a
V_k^{II}	Voting function in district k (k = 1, 2) of middle rank administration II
V_{kMax}^{II}	Maximal votes in district k (k = 1, 2) of middle rank administration II
V^{III}	Voting function in the entire district of top rank administration III
V_k^{III}	Votes out of region k (k = 1, 2)

V_{kN}^{III}	Number of voters in region k (k = 1, 2)
V_{Max}^{III}	Maximal votes in the entire district of the top rank administration III
X	Output of an administration
X_i^I	Output of low rank administration I at location i (i = 1, 2)
X_j^{Ia}	Output of low rank administration I_a at location j (j = 1, 2)
X_k^{II}	Outputs of administrations at low rank I, and middle rank II in region k (k = 1, 2)
$\alpha_{iC}^I, \alpha_{iL}^I$	Output elasticity of low rank administration I at location i (i = 1, 2)
$\alpha_i^I = \alpha_{iC}^I + \alpha_{iL}^I$	Degree of homogeneity of the production function of administration I at location i (i = 1, 2)
$\alpha_{jC}^{Ia}, \alpha_{jL}^{Ia}$	Output elasticity of low rank administration I_a at location j (j = 1, 2)
$\alpha_j^{Ia} = \alpha_{jC}^{Ia} + \alpha_{jL}^{Ia}$	Degree of homogeneity of the production function of the administration I_a at location j (j = 1, 2)
β_i^I	Utility elasticity of the management of low rank administration I at location i (i = 1, 2)
β_j^{Ia}	Utility elasticity of the management of low rank administration I_a at location j (j = 1, 2)
$\delta_{i,k}^{II}$	Abbreviation of parameters referring to the budget adaptation between middle rank administration II in region k (k=1,2) and low rank administration at location i (i = 1, 2)
$\epsilon_{Di,j}^I$	Elasticity on the path of the optimal budget extension of low rank administration I at location i (i = 1, 2) related to the budget of low rank administration I at location j (j = 2, 1)

References

Arnold V. (1992): *Theorie der Kollektivgüter*, Vahlen, München

Batey P.W.J. and Friedrich P. (2000): Aspects of Regional Competition. In: Batey P.W.J. and Friedrich P. (eds.) *Regional Competition*, Springer, Berlin, Heidelberg, New York, pp. 3-33

Berndt C. (1979): *Finanzpolitische Maßnahmen als Mittel zur Beeinflussung von Wahlentscheidungen im föderalistischen Staatswesen*, Dissertation, Siegen

Bös D. (1986): *Public Enterprise Economics*, North-Holland, Amsterdam, New York, Oxford

Bös D. (1989): *Public Enterprise Economics, Theory and Application*, North-Holland, Amsterdam, New York, Oxford

Bös D. (1991): *Privatisation: A Theoretical Treatment*, Calrendon Press, Oxford

Borcherding T. (1977) (ed.): *Budgets and Bureaucrats*, Duke, Durham

Breton A. (1996): *Competitive Governments, An Economic Theory of Politics and Public Finance,* Cambridge University Press, New York

Brezis E.S. and Krugman P.R. (1997): Technology and the Life Cycle of Cities, *Journal of Economic Growth* 2, 369-383

Buhr W. and Friedrich P. (1978): Wichtige Aspekte der Konkurrenz zwischen kleinen Regionen. In: Buhr W. and Friedrich, P. (eds) *Konkurrenz zwischen kleinen Regionen – Competition among Small Regions*, Nomos, Baden-Baden, pp. 11-40

Cheshire P. and Gordon I.R. (1998): Territorial Competition: Some Lessons for Policies. *Annals of Regional Science* 32, 321-346

Christiansen U. (2000): The Concepts of Market and Competition in Political Discourses Involving Geographical Space – History and Some Examples. In: Friedrich P. and Jutila S. (eds.) *Policies of Regional Competition*, Nomos, Baden-Baden, pp. 62-79

Dohse D. (2000): *Regionen als Innovationsmotoren: Zur Neuorientierung in der deutschen Technologiepolitik*, Discussion Papers 366, Institut für Weltwirtschaft, Kiel

Eichhorn P. (1978): *Public Enterprise in the EEC*, Sijthoff &Noordhoff, Aalphen aan den Rijn

Eichhorn P. (1997): *Öffentliche Betriebswirtschaftslehre*, Nomos, Baden-Baden

Eichhorn P. and Friedrich, P. (1976): *Verwaltungsökonomie I*, Nomos, Baden-Baden

Epple D. and Zelenitz A. (1981): The Implications of Competition among Jurisdictions, *Journal of Political Economy* 89, 1197-1217

Feng X. (1998): *Die chinesische 'Treuhandanstalt' – Ein Instrument zur Umwandlung des Wirtschaftssystems in der Volksrepublik China*, Nomos, Baden-Baden

Frey R. (1977): *Zwischen Föderalismus und Zentralismus*, Lang, Bern, Frankfurt

Frey B. (1997): *Ein neuer Föderalismus für Europa: Die Idee der FOCJ*, Mohr, Tübingen

Frey B. (1999): *The New Democratic Federalism for Europe*, Elgar, Cheltenham

Friedrich P. (1969): *Volkswirtschaftliche Investitionskriterien für Gemeindeunternehmen*, Mohr, Tübingen

Friedrich P. (1976): *Standorttheorie für öffentliche Verwaltungen*, Nomos, Baden-Baden

Friedrich P. (1977): Modelle kommunalen Industrieansiedlungswettbewerbs. In: Funck R. (ed.) *Über Wachstum und Wachstumslenkung*, Universität Karlsruhe, Karlsruhe, pp. 22-90

Friedrich P. (1984): Regional Aspects of X-Inefficiency in the Public Sector. In: Wille E. (ed.) *Konzeptionelle Probleme öffentlicher Planung*, Lang, Bern, pp. 189-244

Friedrich P. (1987): Regional Competition under Stagnation. In: Friedrich P. and van Rompuy (eds.) *Fiscal Decentralization*, Nomos, Baden-Baden, pp. 84-133

Friedrich P. (1992): Der vertikale Wettbewerb im Prinzipal-Agent-Verhältnis: Ein Ansatz zur Weiterentwicklung der Theorie der öffentlichen Unternehmen, *Zeitschrift für öffentliche und gemeinnützige Unternehmen*, Beiheft 14, 118-207

Friedrich P. and Feng, X. (1995): Ansätze einer Theorie des Vertragsmanagements der Treuhandanstalt, *Zeitschrift für öffentliche und gemeinnützige Unternehmen* 18, 277-289

Friedrich P. and Feng, X. (1998): Regional Aspects of Treuhand Contract Management. In: Westeren K. I. (ed.) *Cross Border Cooperation and Strategies for Development in Peripheral Areas*. North-Trondelag Research Institute, Steinkjer, pp, 471-508

Friedrich P. and Feng, X. (2000): The Role of Public Enterprises in Regional Competition. In: Batey P.W.J. and Friedrich P. (eds.): *Regional Competition*, Springer, Berlin, Heidelberg, New York, pp. 186-201

Friedrich P. and Jutila S. (2000): Aspects of Policies of Regional Competition. In: Friedrich P. and Jutila S. (eds.) *Policies of Regional Competition*, Nomos, Baden-Baden, pp. 13-60

Friedrich P. and Lindemann S. (2000): A Two-levelled Approach to Municipal in Business Promotion. In: Bojar E. (ed.): *Competition and Co-existence in the Process of European Integration*, Polish Scientific Publishers, Warsaw, pp. 99-125

Friedrich P. and Pfeilsticker A. (1985): Bestimmung der Zahl öffentlich Bediensteter. In: Häuser K. (ed.) *Produktivitätsentwicklung staatlicher Leistungen*, Duncker & Humblot, Berlin, pp. 199-226

Funck R. (2000): Hard and Soft Determinants of Interregional Competition. In: Batey P.W.J. and Friedrich P. (eds.) *Regional Competition*, Springer, Berlin, Heidelberg, New York, pp. 66-86

Gerken L. (ed.) (1995): *Competition among Institutions*, Macmillan, Basingstoke

Gerken L. (1999): *Der Wettbewerb der Staaten*, Mohr, Tübingen

Goudriaan R. and Groot H. de (1991) A Principal-Agent Model or Regulation Applied to the Case of American Universities. In: Prud'homme R. (ed.) *Public Finance with Several Level of Government,* Foundation Journal Public Finance, Königstein, pp. 181-194

Hewings G.J.D. and Sonis M. (2000), Regional Competition and Complementarity: Comparitive Advantages/Disadvantages and Increasing/Diminishing Returns in Discrete Relative Dynamics. In: Batey P.W.J. and Friedrich P. (eds.) *Regional Competition*, Springer, Berlin, Heidelberg, New York, pp. 139-158

Hirsch H. (1992): Hans Ritschls Theorie der öffentlichen Unternehmungen als Element einer Theorie der Staatswirtschaft im Vergleich mit den von Dieter Bös und Blankart vorgelegten Theorien. In: Friedrich P. (ed.) Beiträge zur Theorie öffentlicher Unternehmen, *Zeitschrift für öffentliche und gemeinwirtschaftliche Unternehmen*, Beiheft 14, 20-44

Hirschman A.V. (1958): *The Strategy of Economic Development*, Yale University Press, New Haven [Conn]

Kirsch G. (ed.) (1977): *Föderalismus*, Fischer, Stuttgart, New York

Krelle W. (1976): *Preistheorie*, Teil II, Mohr, Tübingen

Kühne K. (1992): Firmentheorie und Managertum, *Zeitschrift für öffentliche und gemeinwirtschaftliche Unternehmen*, Beiheft 14, 157-177

Lindemann S. (1999): *Theorie und Empirie kommunalen Wirtschaftsförderwettbewerbs*, Nomos, Baden-Baden

Müller S. and Kornmeier M. (2000): *Internationale Wettbewerbsfähigkeit*, Vahlen, München

Niskanen W.A. (1971): *Bureaucracy and Representative Government*, Aldine-Atherton, Chicago

Oates W.E. (1991): Fiscal Federalism: Overview. In: Prud'homme R. (ed.) *Public Finance with Several Level of Government,* Foundation Journal Public Finance, Königstein, pp. 1-18

Oettle K (1976): *Grundfragen öffentlicher Betriebe I, II*, Nomos, Baden-Baden

Parr J.B. (2000): Notes on the Regional Dimension of Rostow's Theory of Growth, Discussion Paper, University of Glasgow, Glasgow

Peltzman S. (1971): Pricing in Public and Private Enterprises: Electric Utilities in the United States, *Journal of Law and Economics* 14, 109-147

Poot J. (2000): Reflection on Local and Economy-Wide Effects of Territorial Competition. In: Batey P. and Friedrich P. (eds.) *Regional Competition*, Springer, Berlin, Heidelberg, New York, pp. 205-230

Postlep, R.D. (1993): *Gesamtwirtschaftliche Analyse kommunaler Finanzpolitik*, Nomos, Baden-Baden

Putnoki H. (1990): *Optimale Wasserpreise,* Nomos, Baden-Baden

Rees R. (1993): *The Economics of Regulation and Public Enterprises*, Harvester Wheatsheaf, Oxford

Renault P. and Winden F. van (1991): A Theoretical Model of a Multi-Level Government Applied to the Netherlands. In: Prud'homme R. (ed.) *Public Finance with Several Level of Government,* Foundation Journal Public Finance, Königstein, pp. 263-277

Ritschl H. (1931): *Gemeinwirtschaft und kapitalistische Wirtschaft*, Mohr, Tübingen 1931

Rompuy P. van (1977): A Fiscal Policy Model in a Two-Region Monetary Model, *Regional Science Research Paper* 18, Centrum voor Economische Studien, Leuven

Rostow W.W. (1960): *The Stages of Economic Growth: A Non-Communist Manifesto*, Cambridge University Press, Cambridge
Schwager R. (2000): Nimby Facilities and Metropolitan Governments, *Jahrbuch für Regionalwissenschaft* 20, 39-54
Thiemeyer T. (1970): *Gemeinwirtschaftlichkeit als Ordnungsprinzip*, Duncker & Humblot, Berlin
Thiemeyer T. (1975): *Wirtschaftslehre öffentlicher Betriebe*, Rohwohlt, Reinbek
Thöni E. (1986): *Politökonomische Theorie des Föderalismus*, Nomos, Baden-Baden
Wildasin D.E. (1986): *Urban Public Finance*, Harwood, London
Wonnemann H. G. (1989): *Budgeteffekte der Standortwahl von Verwaltungen*, Nomos, Baden-Baden
Ziemes G. (1992): Politik-ökonomische Betrachtungen öffentlicher Unternehmen, *Zeitschrift für öffentliche und gemeinwirtschaftliche Unternehmen*, Beiheft 14, 56-82
Zimmermann H. (1999): *Kommunalfinanzen*, Nomos, Baden-Baden

6 Regional Cooperation of the Baltic States: Implications for Membership in the European Union

*Manas Chatterji * and Bengt Lorendahl ***
* Binghamton University, School of Management, and
Indian Institute of Management, Calcutta, India
** Sodertorns University, Sweden

6.1 Introduction

Although the economic, social and political structures of many countries in the world are increasingly moving to a global pattern, there is still an immense disparity in the living conditions between poor and rich countries. All countries begin with an agrarian background and gradually move to manufacturing, services and eventually to a knowledge-based society. Many developing countries are lagging far behind. The rates of population growth in these countries are still substantial particularly due to steady birth rates and declining death rates. Agriculture is still prominent in many of these countries and at the same time there has been an explosion in urban populations with its associated problems. Energy has played a major role in the development process of rich countries. However, beginning in the sixties, it was found that excessive use of energy affected the environment and human health. It has now been realised that the availability of water will be the primary concern of the 21st century. Simultaneously, political systems are moving away from dictatorial and socialist economic systems to more democratic and capitalistic structures. There has also been a significant movement towards human rights and free trade.

Due to limitations of dwindling resources, the pressures of economic inequality and the inevitable rise in individual expectations, conflict will play a pivotal role in the 21st century. Since it is no longer possible to have a utopian view, it is urgent that we face this challenge of conflict for the developing, developed and emerging countries. Cooperation between countries and regions within a country is essential for conflict management and peaceful coexistence. The objective of this contribution is to develop some comparative theoretical models to accurately characterise these resource based conflicts. This chapter will specifically focus on conflict management in Baltic States countries. The discussion will also be quite relevant to characterise the relationship between the territory of the United States and Canada which borders the Great Lakes and elsewhere.

6.2 Background of the Baltic States

Definitions and demarcations: It is not often observed that there is a difference between the two concepts of the Baltic States and the Baltic Sea States. The first one usually refers to the three countries Estonia, Latvia and Lithuania, sometimes together as an entity they are called Balticum. The second one comprises a more extensive geographical area, also referred to as the Baltic Sea Region [BSR]. There is not yet an unambiguous, precise and generally recognised definition of the BSR area, as will be shown below. It is very important, especially in studies of the three small Baltic States, to simultaneously pay attention to the wider context of BSR. A great deal of cooperative action for development in these countries is carried out under the auspices of the organised cooperation of all the Baltic Sea States.

If we ask anyone what they regard as the Baltic Sea Region this question would probably generate answers like the nations surrounding the Baltic Sea or the coastal areas of these countries. In fact, current definitions to some extent combine these two responses. This characterisation illustrates the necessity of cooperative action. Eleven nations are engaged in such top-level cooperation at the national government level. The members of this *Council of the Baltic Sea States* [*CBSS*] include: Denmark, Estonia, Finland, Germany, Iceland, Latvia, Lithuania, Norway, Poland, Russia and Sweden. These are then the official Baltic Sea States.

If the national borders of these states are supposed to signify the extension of the BSR area, then this is an unambiguous and imprecise delimitation. It lacks realism – thus defined the Baltic Sea region would extend from the Arctic Ocean to the Swiss border, and from Reykjavik to Vladivostok! As for Russia, this vast country now has very small districts adjoining the Baltic Sea – but they indisputably have *some* Baltic Sea coast. Perhaps more striking is the fact that two countries – Iceland and Norway – which have no coastal areas at all by the Baltic Sea are included and defined as Baltic Sea States! Undoubtedly, this is the result of the long tradition of cooperation between all the five Nordic countries.

In other contexts, the BSR is delimited somewhat differently. In the guiding principles for a research fund at University of Södertörn, *The Baltic Sea Foundation,* both nations and parts of nations are specified. The geographical extension is limited westward [Iceland and Norway are excluded], eastward [to Northwestern Russia] and southward [to areas in Northern Germany].

In short, the CBSS delimitation offers an unambiguous but unrealistic definition of the Baltic Sea Region. The research foundation's definition is more realistic but somewhat imprecise. A third variant in defining the BSR seems to combine realism and precision. In a study by the Baltic Ring [a cooperative project between 18 power companies and utility organisations from the countries around the Baltic Sea for developing a common electricity market] a distinction is made between countries directly bordering the Baltic Sea [Denmark, Estonia, Finland, Germany, Latvia, Lithuania, Poland, Russia and Sweden]. This definition is supplemented by two countries [Belarus and Norway] which are characterised by vicinity to the Baltic Sea. The *Länder* and cities are specified for Germany, the *Oblasts* and republics are included for Russia.

These unambiguous geographical borders drawn around BSR in the Baltic Ring Study make it possible to present a map over the region, as well as assemble data on the population and its distribution. In this delimitation, the total population of the region is about 127 million people. Poland [30.3 percent] and [northern] Germany [28.6 percent] heavily dominate the population distribution. None of the other countries have a share that exceeds 10 percent. The smallest percentage shares are found in the three Baltic states: Estonia [1.2 percent], Latvia [2.1 percent] and Lithuania [2.9 percent].

Re-integration – The importance of history

In two reports analysing cooperation in European frontier regions, border problems and possible obstacles to border-crossing cooperation have been discussed. Some of these problems and obstacles primarily concern subnational frontier regions, but some are also highly relevant to interstate cooperation. In this context, one should pay attention to the importance of non-economic factors in border-crossing cooperation; obstacles [or possibilities] rooted in historical, political and psychological factors that must be uncovered. Most neighbouring regions and nations have a common political past. Transborder cooperation is easier in these areas where this common past is more distant and where there have been no military or cultural conflicts in recent times.

A common history: Shared memories, and often a shared culture are obviously of great importance for progress in transborder regional cooperation. Even if this background seems more relevant in directly adjacent frontier regions for historical and cultural reasons it is also important for all countries around the Baltic Sea. Common history is a potential ground for successful cooperation in countries and regions which have been integrated or unified in earlier periods of history – provided that the relations have not been characterised by oppression and exploitation.

Large parts of the Baltic Sea Region have been integrated or unified for considerable periods of history – which warrants the term re-integration used in this discussion. Two such periods and states of earlier integration will be briefly described.

The Hanseatic League

Although originally founded in London in the 12th century, by German merchants, the Hanseatic League is primarily associated with a number of cities around the Baltic Sea, two of the most prominent being Lübeck and Visby. It encompassed those cities which had an interest in Baltic Sea trade. The Hanseates attained its strongest power position in the 14th century when 70 cities belonged to the league. These cities constituted a free association without a firm organisation. They were capable of joint efforts including military action. This made them a powerful force in the Baltic Sea area.

The decline of the Hanseatic League was partly due to the discovery of new trade routes, but also to political circumstances. The new Scandinavian union between Denmark-Norway and Sweden [at that time including Finland] carried on an anti-Hanseatic policy, and so did the Dutch cities outside the league. Gradually, the Hanseatic League was dissolved by the end of the 16th century and the beginning of the 17th century.

Economic objectives, primarily commerce and shipping, dominated this early and long-lasting integration process in the Baltic Sea Region. These activities were combined with or even became the rationale for political and military endeavors. It should also be noted that certain rules of international law were introduced in this period of Hanseatic hegemony. International trade gained increased security and piracy was thwarted in Nordic waters.

The Swedish Baltic Sea Empire

During the same century as the mighty Hanseatic League finally disappeared as a dominant force in the Baltic Sea area, the Swedish Kingdom rose to a position of a great power in Northern Europe. Centuries before, Sweden encompassed land on the Eastern side of the Baltic Sea [Finland and provinces that today belong to Estonia, Latvia and Russia]. After victories in European wars, additional coastal provinces south of the Baltic Sea were incorporated in the Swedish empire [Vorpommern, Stettin, Wismar, Bremen and Verden]. Sweden attained its largest extension in 1658, after encompassing the island of Bornholm and the Norwegian county of Trondheim.

The political map in those years explains the formulation in classic Swedish history books that the Baltic Sea was in practice, a Swedish domestic sea. Historically, the Baltic Sea Region has never since been integrated by a single nation to this extent.

In the following centuries, Sweden gradually lost its overseas provinces, the gravest loss of land being the cession of Finland to Russia in 1809. If we take into account what was indicated above about the importance of shared history for current cooperation, it is especially relevant to note that none of the overseas provinces were separated from the Swedish Kingdom as a result of a rebellion or a liberation movement in the province. This was not the case with Norway, which was in the union up until 1905. All Swedish provinces on the eastern and southern coasts of the Baltic Sea were ceded to other nations after their wartime defeats. Only in two instances was it a matter involving trade.

The Baltic Sea and the Cold War

World War II engendered military confrontation between the Baltic Sea states and in its wake left a restructured map, mainly characterised by the expansion of Soviet Union territory around the sea. Estonia, Latvia and Lithuania became Soviet republics. Finland entered a position of dependency on their eastern neighbour. The southern coast of the sea now belonged to Soviet allies, Poland and GDR. The rest of the Baltic Sea coast became parts of the neutral Sweden and the two NATO countries Denmark and Western Germany. Political tensions and incidents [also involving Sweden] were substantial features of the postwar period before the 90's. The collapse of the Soviet empire became a prerequisite for the transition of the Baltic Sea into a sea of peace. It also became a catalyst for the recently established practice of regional cooperation and re-integration.

Present political objectives

For the countries participating in the current BSR cooperation, the possibility of developing their own economies is a strong driving force. The rapid and steady growth of this regional cooperation clearly indicates their belief that these national and economic objectives can be better attained through joint action. The Baltic Sea States also [not unlike the Hanseatic League] engage in other common activities that facilitate and support trade and economic expansion. These activities are also carried on for their own intrinsic value.

Current examples of cooperation include programs and projects of interest for all parties, such as joint security and environmental activities. Some aim directly at fortifying industry and infrastructure in economies in transition, mainly in the three Baltic States, i.e. Estonia, Latvia and Lithuania. However, it is also obvious that the BSR countries which have a more developed market economy and industrial technology see the potential benefits of this regional cooperation, both for the region as a whole and for their own economic development. In a Swedish Government bill (Hellenburg et al., 1997/98, p. 7), one of the headlines is '*The Baltic Sea Region as our domestic market*'. In the same bill, the Swedish government writes: '*... By the present development, favourable conditions are created for the realisation of a dynamic Northern European region ...*' (Hellenburg et al., 1997/98, p. 9). '*The countries around the Baltic Sea have an inherent dynamics and potential as one of Europe's growth centres ...*'. The government intends to continuously work for the establishing of contacts across the Baltic Sea, for increased trade and investments and for measures which promote knowledge of language, history and culture' (Hellenburg et al., 1997/98, pp. 10-11).

Although focusing on their own region, the Baltic Sea states simultaneously maintain a pronounced European perspective. The Swedish government states [in the previously mentioned bill] that the prospects of regional cooperation can be fully exploited in an integrated Europe. It also recognises the aspirations to become EU members: '*The development cooperation shall be directed towards supporting the membership of Estonia, Latvia, Lithuania and Poland in the European Union ...*' (Hellenburg et al., 1997/98, p. 57).

Fields of regional cooperation

Such diverse examples of BSR cooperation can, be grouped under four main headings or policy spheres, as in the previously cited Government bill. They will be very briefly presented below and illustrated by a few examples.

Security community: In the sphere of security, cooperative efforts will be concentrated upon improving competence in security policy and peace-promoting operations, developing defense organisations under democratic control. Sustained efforts will also be made to support the capability of Baltic countries to control and guard their borders, combat international and organised crime, build up competence in policies of asylum and migration and improve readiness to cope with accidents and catastrophes. Some concrete examples include the establishment of an Institute for Foreign Policy in Riga, Latvia, training of Baltic officers in Swedish military universities and the formation of a special Baltic battalion [BALTBAT] for peace-maintaining operations.

The culture of democracy: The most important and fundamental change among the states in the BSR is the democratisation process. Although the process into democracy has advanced unequally in the countries concerned, for the first time in many decades there are now valuable opportunities to create convivial, neighbourly relations, which are based upon tenets on of mutual respect for democratic values.

Cooperation, not only between states, but also between municipalities, cities and regions is presumed to increase in the region. Intensified academic and cultural cooperation are natural steps in the transition to ordinary neighbour-country relations. In addition to public institutions, other types of organisations, such as entrepreneurial groups, church congregations, and assorted popular movements are now engaged in creating new contacts and networks.

Two concrete measures following this direction include the already established Business Graduate School and the planned Riga Graduate School of Law in Latvia. Two large scholarship programs have also been designed to foster academic cooperation between Sweden and neighbouring countries.

A socially sustainable economic transformation: The primary objectives of such pronounced cooperation include the developing towards a social market economy, restructuring and developing of a viable industry, supporting public economic institutions and eliminating obstacles to trade and investments in the Baltic Sea Region. Examples of concrete measures within this sphere are two programs which support Swedish firms interested in Eastern trade. Under the auspices of *Start-Öst,* credits are available for Swedish SMEs entering into business projects with local firms in cooperating countries. Up until now, more than 200 projects have received loans. Another program, *Advantage East,* supports firms in Estonia, Latvia, Lithuania and the Ukraine striving for increased exports to Sweden. This program also includes business training. In a large-scale business training program – also named the Jeltsin program – 1,000 young Russian business executives will participate in educational programs and practice in Sweden over a five year period.

An environmentally sustainable development: In the appropriately named *Action Program for the Baltic Sea,* the 132 most significant sources of pollution have been identified. Measures have been taken to improve the conditions at most of these 132 hot spots, which were already identified by 1992. The Chernobyl disaster of 1986 was a startling demonstration of the vulnerability of our common environment. Consequently intensive efforts have been made to improve security in the nuclear power stations in BSR and to develop protection against radiation. In the first stages of cooperation involving energy and environment, efforts were concentrated on the nuclear power station Ignalina in Lithuania. Some other examples of concrete and current measures in the environmental field include *Agenda 21 for the Baltic Sea Region*, seven joint water and sewage projects and the previously mentioned Baltic Ring in the energy sector.

6.3 Organisational Structures for Regional Cooperation

The *Council of the Baltic Sea States* [*CBSS*] was established in March 1992, on a Danish-German initiative. Members include the eleven nations enumerated above plus the European Commission. The Council shall act as a forum for guidance and coordination in the majority of questions of common interest to member countries. However, issues of a purely security policy nature are beyond the scope of the Council's mandate.

CBSS was created as an informal cooperation council, in this respect it resembles the Hanseatic League, especially for the foreign ministers of member countries! The foreign minister of the presiding country is responsible for coordinating the Council's on-going activities between Ministerial Sessions, and is assisted in this work by a special *Committee of Senior Officials* [*CSO*].

Some steps have been taken to build a more elaborate and cohesive cooperation structure. In 1996, the Heads of Government met in Visby [the old

Hanseatic city on the island Gotland, in the middle of the Baltic Sea] for the first *Baltic Sea States Summit.* One of the prime ministers served as a coordinator and assumed responsibility for top-level cooperation within the Prime Ministers office. This is how the *The Baltic Sea States Support Group* was established. The second Baltic Sea States Summit took place in Riga in 1998. In addition to the Heads of Government, the President of the European Commission and the President of the European Council participated in the meeting. In autumn of 1998, the BSR top-level cooperation structure increased its organisational continuity and resources. A permanent *CBSS Secretariat* was established in Stockholm.

In addition to the state and top-level structures described above, there are other organisations which have been designed to promote BSR cooperation in various fields of action. One is the *Baltic Sea Fund,* financing projects that promote cooperation and development in five priority sectors: food-stuffs, energy, education, infrastructure and environment. The most prominent business operators in the Baltic Sea countries also gather in summit meetings in *Baltic Sea Business Summit.* In addition to the institutions previously mentioned above, and in earlier sections, a number of other public authorities and non-governmental organisations participate and are assigned roles in the organisational framework of the Baltic Sea Region cooperation.

6.4 Contemporary Situation in Baltic States

Estonia, Latvia and Lithuania are referred to as Baltic States. They were part of the former Soviet Union. They declared independence in the nineties after 50 years of annexation by the Soviet Union. Table 6.1 provides some information about these countries. It is clear from the above table that although the population of these countries is quite small, the per capita income is quite high. The linkages of these countries to the rest of the world is much higher compared to other former Soviet republics. As is seen from Table 6.2, the population density is much lower compared to those of some other European countries and the life expectancy level is on a par with Finland.

The GDP per capita figures are not reliable. Table 6.2 also provides information about minority populations. It can be seen from the table that there is a substantial Russian minority population in Estonia and Latvia, compared to Lithuania. This is an important factor for economic transformation for these countries. The fixed exchange rate of those countries as seen from Table 6.3 seems to be stable. Although the data on index of industrial production in Table 6.4 is very outdated, it gives us an idea about the changes over time.

Table 6.1 Soviet Union: Population, national income and share of trade by republic

	Population [% of total in 1989]	National Income [Net Material Product, NMP]			NMP per Capita [Roubles, 1989]	Percent of Average NMP per capita	Deliveries to Other Republics	Exports Abroad
		Total	Industry	Agriculture			[% of NMP in Current Domestic Prices]	
		[Current Prices, % of total in 1988]						
Estonia	0.5	0.6	0.6	0.7	2,780	119	66.5	7.4
Larvia	0.9	1.1	1.1	1.2	2,844	121	64.1	5.7
Lithuania	1.3	1.4	1.1	1.9	2,584	110	60.9	5.9
Armenja	1.1	0.9	1.2	0.7	2,102	90	63.7	1.4
Azerbaijan	2.5	1.7	1.7	2.2	1,621	69	58.7	3.7
Belorussia	3.6	4.2	4.0	4.9	2,764	118	69.6	6.6
Georgia	1.9	1.6	1.4	2.1	1,861	79	53.7	3.9
Kazakhstan	5.8	4.3	2.5	6.1	1,713	73	30.9	3.0
Kirghizia	1.5	0.8	0.6	1.3	1,297	55	50.2	1.2
Moldavia	1.5	1.2	1.0	1.8	1,968	84	62.1	3.4
Russia	51.4	61.1	61.9	18.0	2,794	119	18,0	8,6
Tadjikistan	1.8	0.8	0.5	1.2	917	39	41.8	6.9
Turkmenistan	1.2	0.7	0.4	1.2	1,363	58	50.7	4.2
Ukraine	18.0	16.3	16.7	17.1	2,104	90	39.1	6.7
Uzbekistan	6.9	3.3	2.3	5.2	1,069	46	43.2	7.4
Total	100.0	100.0	100.0	100.0	–	–	–	–

Source: Goskomstat (1990)

Table 6.2 Basic figures for the Baltic States

	Estonia		Latvia		Lithuania	
GDP, 1991 [US$ per capita]	3,830		3,410		2,710	
Area [km²]	45,215		64,600		65,200	
Population, 1990 [thousands]	1,575		2,670		3,737	
Population Density, 1990 [per km²]	35		40		57	
Birth Rate	14.2		14.2		15.1	
Death Rate	12.4		13.0		10.6	
Life Expectancy at Birth [years]	70.8		70.5		72.0	
Share of Titular Nationality [%]	61.5		52.0		79.6	
Share of National Minorities [%]	Russians	30.3	Russians	34.0	Russians	9.4
	Ukrainians	3.1	Belorussians	4.5	Poles	7.0
	Belorussian	1.8	Ukrainians	3.5	Belorussians	1.7

Source: World Bank (1991)

There have been drastic changes already in tax laws, particularly in the development of small and medium-sized enterprises. It is slow in Estonia. Latvia has stipulated a 20 percent tax concession for small enterprises. The same is also now true for Lithuania in small enterprises. The Estonian Business Code (September 1995) encourages entrepreneurs in cooperative societies, foreign investment, insurance. Estonian law places no general restrictions on foreign investment. In Latvia foreign investors can also do business with or without local participation. Budget and tax reforms work together to promote economic and social transformation. In Estonia (1992) budgeting was decentralised and an independent social security budget was created in Latvia (1991). Tax laws adopted in Baltic States determined specific types of taxes and duties, including: personal income tax, enterprise income tax, withholding tax, value added tax [VAT], excise duties, the social tax, etc. For state tax revenues the most important taxes are income taxes, VAT, excise tax and social tax. Estonia has used a simple and proportional tax rate.

Table 6.3 USD exchange rates in the Baltic States in 1992-1994

Year	Estonia	Latvia	Lithuania
1992	11.84	0.683	3.79
1993	13.22	0.674	3.90
I	12.94	0.842	3.96
II	12.12	0.819	4.43
III	13.19	0.714	4.88
IV	12.78	0.671	5.35
V	12.84	0.662	5.09
VI	13.15	0.648	4.50
VII	13.70	0.652	4.15
VIII	13.59	0.639	4.00
IX	13.10	0.621	4.25
X	13.10	0.609	3.96
XI	13.59	0.608	3.93
XII	13.67	0.598	3.90
1994	13.40	0.560	4.00
I	13.94	0.591	3.90
II	13.89	0.577	3.90
III	13.54	0.570	3.93
IV	13.58	0.569	4.00
V	13.26	0.565	4.00
VI	13.06	0.561	4.00
VII	12.56	0.550	4.00
VIII	12.51	0.549	4.00
IX	12.41	0.547	4.00
X	12.17	0.542	4.00
XI	12.31	0.543	4.00
XII	12.57	0.551	4.00

Sources: Government of Latvia (1994, 1994a, 1995, 1995a)

There have been many changes in Baltic States since we started writing the paper. Let us update. Latvia has closed 18 chapters of negotiation and 29 are still open. (October 2001). Many observers believe that the new countries will become full member of the EU in time for the election of the European Parliament in 2004. Since present economic conditions in Lithuania are not promising and the process of economic reform is going slowly, the invitation is a promissory note which must be paid in the future by restructuring the economy. Latvia is making important steps toward such reform in the areas of education, agriculture, labour, health care. Lithuania has to reduce the state subsidy drastically. Estonia was invited before the other two countries. There was some dissatisfaction. The situation is normalising and the three countries now realise that their secure and successful future lies in mutual cooperation and not in selfish isolation. Estonia was trying to portray itself as a Nordic country and group Lithuania with Poland and Central European Countries and is placing Latvia somewhere in between. This attitude needs to be changed. Also the position of Russia vis a vis NATO and EU is changing.

Table 6.4 Industrial production (1940 = 1) in the Baltic States

	Estonia	Latvia	Lithuania
1950	1.4	1.3	0.8
1960	4.8	4.6	4.3
1970	11.6	11.4	13.0
1980	20.1	19.0	24.0

Russia is moving toward a tighter economic integration with the EU, in order to solve its economic development. Moreover, the problems of Russian minorities in Baltic States could hardly be related to the EU enlargement and should be solved on a bilateral basis by the States involved. Logically, Estonia should stress its simultaneous membership in both groups (NATO and EU). EU expansion is based on political criteria. They would rather divide the relationship with EU on MFN (most favoured nation) principles rather than with the Baltic States. However, it is Russia's intent to keep a good relationship with EU and it is not expected that it will push this point to the extreme. It is estimated that when Baltic States become members, about half of Russia's trade will be with the EU. After developing closer economic ties with Russia and deepening integration with EU, the Baltic States will make connections with the East as well as the West. Such a move is impossible without close cooperation between the Baltic States. The enormous power anomalies between the three Baltic States determine the need for Baltic integration in NATO. Strong buffer states can serve Russian interests better than weaker states. Baltic States pay hard currency for Russian gas. In the past, heterogeneous security interests needed limited cooperation. However, the events of the past few years imply that more cooperation is needed for security. This is the reason why countries like US, Belarus and Ukraine are all involved in Baltic Sea projects. Enlargement of EU is the most single event to stabilise and increase the security. Accession implies both security benefits and greater obligations.

Latvia has used a differential tax rate, whereas Lithuania has a complicated tax rate schedule. A concrete social security system with respect to pension insurance, children benefits, medical and sickness benefits has also been implemented.

Table 6.5 provides some detailed information. Although information about trade given in Table 6.6 is not reliable, it can be said that the countries have a high dependency on energy import. In industrial production, it has strong dependency on metallurgy and machinery. The region is a net exporter of light industry products like fabrics, clothing and leatherwear. Agricultural products play a very minor role. Table 6.6 gives some information in this respect. It is interesting to ask that if these countries had been integrated with the rest of Europe or Western Europe today, what would the trade pattern have been? This can be estimated by using a gravity model of the type:

$$X_{ij} = b_0 + b_1\,(GDP_j/POP_j) + b_2\,GDP_j + b_3\,(GDP_i/POP_i) + b_4\,GDP_i + b_5\,DIST_{ij} + \text{dummies}$$

where X_{ij} denotes total exports from country i to country j, GDP is the total Gross Domestic Product and POP is the total population of the relevant countries and DIST the physical distance of the relevant countries. In addition, there is a series of dummy variables for specific features of the countries and the data sets.

Table 6.5 Taxation in the Baltic States

		Estonia	Latvia	Lithuania
Enterprise Income Tax	1992	15%-23%-30%	25%	29%
		35%[1]	35%; 45% 65[2]	19%[3]
	1993	35%	25%	29%
			35%; 45% 65%[2]	19%
	1994	26%	25%	29%
			35%; 45% 65%[2]	19%
Personal Income Tax	1992	16%-24% - 33%	16%[4]	18% - 33%
	1993	16%-24% - 33%	16%[4]	18% - 33%
	1994	26%	25%[4]	33%
Social Tax	1992	20% + 13% employers[5]	37% employers + 1% employees; 19% farmers	30% employers + 1% employees 23% farmers
	1993	20% + 13% employers	37% employers + 1% employees; 19% farmers	30% employers + 1% employees 23% farmers
	1994	20% + 13% employers	37% employers + 1% employees; 19% farmers	30% employers + 1% employees 23% farmers
Value Added Class	1992	10%	6% food, 12%	18%
		18%	other goods	
	1993	18%	19% food, 18% other goods	18%
	1994	18%	18% all goods [6]	18%

Notes: 1 since June 1992; 2 depending on the field of activity; 3 farmers; 4 since April 1, 1994; 5 20% social tax and 13% health insurance tax; 6 since April 1, 1994

Although the use of Gravity Models to analyse spatial interactions started in the fifties and sixties, mainly in the area of Geography and Regional Science, in recent years the model has been extended to Economics and International Trade. There has been tremendous development in the estimation procedure of the parameters of the model. Fischer et al. (1999) present an excellent theoretical and empirical work on the use of a global search procedure proposed by Storn and Price (1997). This approach helps us to obtain a global optimum. In Fischer and Getis (1999), readers will get excellent related studies.

Take a specific country, for example, Latvia. The relationship of foreign trade and GDP shows that the country is becoming more open but has a deficit trade. It relies on low value added products which signify the low competitiveness of local producers. The major investors in Latvia are Finnish Telcom, Polorbak, Shell Overseas. Others invest in the service sector, food industry, energy and wood processing. Table 6.7 gives some current information of foreign trade of Latvia with different countries.

Table 6.6 Composition of trade in world prices, 1989 [million roubles]

	Estonia		Latvia		Lithuania	
	Export	Import	Export	Import	Export	Import
Industry	1,861	3,262	3,995	5,375	4,530	7,608
of which %						
Energy	*10*	*17*	*3*	*19*	*24*	*33*
Metallurgy and Machinery	*33*	*52*	*53*	*54*	*45*	*50*
Chemical and Petroleum	*13*	*13*	*13*	*9*	*7*	*8*
Wood, Paper, Building Materials	*6*	*3*	*4*	*4*	*4*	*2*
Light Industry	*17*	*7*	*7*	*4*	*10*	*4*
Food Production	*18*	*5*	*15*	*7*	*10*	*4*
Other Industries	*3*	*2*	*4*	*2*	*0*	*0*
Agriculture	17	76	47	188	65	197
Other Production Sectors	190	125	371	78	224	286
Total	2,068	3,462	4,413	5,640	4,819	8,091

Source: World Bank (1989)

Although the three countries are politically independent, the real test of viability once depended upon economic independence and integration with the European community. This is by no means the case now. The future greatly depends on the success of the transformational process, national economic sovereignty, transforming the structure of production, enterprise reform, impact of economic reform on household, macroeconomic issues, stabilisation, liberalisation and privatisation. Another important issue is security, however security issues and economic transformation are not the topic of discussion here. Our goal is to determine how the country can cooperate, and how the mutual cooperation of these countries and other Baltic Sea countries [Sweden, Denmark, Germany, etc.] can hasten this integration. What is the real role of EU?

Table 6.7 Foreign trade turnover by country (I-VI 1998) [in Latvia currency]

Countries	Animals		Various Industrial Goods		Chemicals		Metals		Machines		Food		Textiles		Wood	
	Export	Import	Export	Import	Export	Import	Export	Import	Export	Import	Export	Import	Export	Import	Export	Import
Total	16.1	19.8	25.6	20.3	33.4	190	52.6	76.4	39.9	177.3	47.4	69.4	87.5	70.6	184.0	7.7
Russia	5.3	0.6	1.8	0.1	7.4	9.6	4.2	19.3	16.4	6.8	20.2	0.4	11.7	1.9	0.9	1.0
France	2.4	0.2	2.6	0.4	0.1	5.2	0.1	0.7	0.1	3.2	0.0	2.7	2.1	1.4	2.2	0.1
Estonia	2.2	1.7	0.8	1.3	4.5	7.1	3.2	8.0	1.6	7.4	3.2	6.3	2.1	3.5	1.6	0.7
Lithuania	0.9	2.7	0.6	1.4	9.6	6.9	4.9	3.5	2.9	7.4	4.5	6.2	5.2	2.2	2.7	0.6
Norway	0.0	3.0	0.3	0.6	0.1	0.5	0.0	0.5	0.1	3.1	0.0	0.1	1.6	1.8	1.6	0.0
Poland	0.0	1.8	0.5	2.0	0.0	4.3	0.2	2.1	0.5	4.9	0.4	2.3	0.4	2.1	3.2	1.0
Germany	1.0	1.2	6.6	3.6	1.5	17.2	8.0	9.4	7.4	35.2	1.0	7.2	17.9	16.6	29.7	1.2
Denmark	1.0	1.2	5.6	0.8	0.9	2.0	0.3	1.1	0.5	8.2	0.1	2.6	8.8	6.9	5.6	0.0
Italy	0.0	0.0	0.3	2.8	0.0	0.8	1.7	3.2	0.1	10.9	0.0	1.9	4.6	3.9	0.9	0.1
Finland	0.0	0.1	1.2	1.5	0.1	10.2	0.2	5.6	0.8	28.4	0.1	2.8	2.2	2.5	3.3	1.1
U.S.	0.0	1.8	0.2	0.3	0.3	1.9	12.8	0.3	2.9	6.2	0.3	1.4	0.6	0.5	0.8	0.1
Ukraine	0.3	0.0	0.2	0.0	2.3	4.9	0.0	0.0	0.0	0.3	4.3	0.3	2.1	0.1	0.9	0.0
Sweden	0.6	1.0	1.3	1.3	0.3	5.2	0.6	3.2	1.3	13.7	0.0	1.7	9.3	10.9	41.2	0.4
Great Britain	0.0	1.2	1.1	0.5	0.2	4.9	0.4	3.6	0.4	7.4	0.0	2.2	2.4	2.3	65.9	0.1

Source: Central Statistical Bureau of Latvia (1998)

In addition to economic integration, there is an urgent need to cooperate in controlling air and water pollution. The Baltic Sea is highly polluted. After World War II, chemical weapons, German nerve gas, and poison gas were dumped in the Baltic. They became time bombs bound to explode. The U.S., Russia and Britain all dumped this toxic material in the Baltic Sea. The available information on the poisoning of the marine environment and fish by released chemical warfare agents is very limited and is mostly related to laboratory investigations. Although the complex ecosystems of the Baltic Sea have been studied at length, there is a potential threat to marine environments that requires more detailed scientific investigations.

Table 6.8 Water consumption and sewage discharge, 1989 [million cubic meters]

	Estonia	Latvia	Lithuania
Total Extracted Water	3,328	733	4,355
From Underground Sources	456	320	506
Utilised Total	3,014	650	4,327
Total Discharge of Waste Water	3,248	579	4,075
Not Needing Treatment	2,726	212	3,625
Sewage in Need of Treatment	517 (100%)	367% (100%)	450 (100%)
• **Standardly Treated**	271 (52%)	110 (30%)	114 (25%)
• **Semitreated**	192 (38%)	–	124 (28%)
• **Untreated**	54 (10%)	257 (70%)[1]	211 (47%)

Note: 1 Untreated or insufficiently treated

Sources: Estonian Nature Management Scientific Information Centre (1990), Government of Latvia (1990); Government of Lithuania (1990)

In addition, weapon dumping countries bordering the Baltic Sea have indiscriminately polluted the sea. The question that arises for environmental justice is who should be responsible for cleaning up the pollution. Three Baltic States and Russia have been culprits in polluting the water. Baltic countries, with the exception of Northern Estonia, have abundant resources of surface and groundwater. There are three major sources of polluted water: insufficiently treated sewage from towns and municipalities, industrial sewage and agriculture. Estonia and Lithuania are the largest water users, 80 percent of which are used as cooling water for thermal power plants and the remainder is used equally for industry, household and service. Table 6.8 gives water consumption figures. Per capita water consumption varies considerably between the three states. Existing sewage treatment is far from satisfactory. Biological chemical treatment is rare due to a lack of equipment. Table 6.9 provides information about discharge by different cities. Table 6.10 gives data for chemical discharge to the Baltic Sea.

Table 6.9 Discharge of sewage from major baltic towns, 1989 [million cubic meters]

	Total Discharge of Sewage	Standardly Treated	Semitreated	Untreated [in %]
Estonia	527	271	256	100.0
Tallinn	117	3	114	46.3
Kohtla-Jarve	295	237	58	23.6
Harju	19	7	12	4.9
Rakvere	11	5	6	2.4
Narva	28	25	3	1.2
Latvia	367	110	257	100.0
Riga	171	14	157	61.1
Liepaya	22	0	22	8.6
Daugavpils	21	3	18	7.0
Jurmals	22	7	15	5.8
Jelgava	10	7	3	1.1
Lithuania	450	114	336	100.0
Kaunas	89	0	89	26.5
Vilnius	88	0	88	26.2
Klaipeda	46	0	46	13.7
Siauliai	21	11	10	3.0
Panevesys	22	13	9	2.7

Source: Central Statistical Bureau of Latvia (1998)

Table 6.10 Discharge of nutrients, oil, and phenol to the Baltic Sea[1], 1989 [tons]

	Nitrates	Phosphorous	Oil Products	Phenol
Estonia	10,350	588	240	523
Latvia	7,155	944	640	2[2]
Lithuania	7,770	881	540	–
Total	25,275	2,413	1,420	525

Notes: 1 Discharged directly to the sea or transported by rivers; 2 the figure for Latvia is probably underestimated; for Lithuania there is no information available

Source: Central Statistical Bureau of Latvia (1998)

Estonia is the largest emitter of air pollution among the Baltic Republics (see Table 6.11 and Table 6.12). Estonia has chosen an administrative model for pollution control compared to European countries. In Latvia [like Lithuania] the main body which controls pollution is the Environmental Protection Committee. Although the environmental problems in Baltic States are not as catastrophic as in Russia, ecology is in conflict with economic priorities. Several international finding agencies such as the European Bank for Reconstruction and Development and the Nordic Environment Finance Group are financing development projects. The achievement of independence by Baltic States now enables them to deal with maritime border delimitation, treaty devaluation and exclusionary regimes.

Table 6.11 Air pollution in the Baltic States, 1989 [thousand tons]

	Estonia	Latvia	Lithuania
Total Emissions	1,050	585	approx. 1,000
From Traffic	474	426	approx. 570
From Stationary Sources	576	159	431
• Solid Particles	261	38	44
• Sulphur Dioxide [SO_2]	210	59	188
• Nitric Oxides [NO]	23	12	44
Total per Capita [kg]	668	217	270
[SO_2] per Capita [kg]	134	22	51
[SO_2] per km^2 [kg])]	4,656	926	2,883

Note: The figures given for traffic emissions should be regarded as rough estimates.
Source: Central Statistical Bureau of Latvia (1998)

Table 6.12 Air pollution from the most polluting towns, 1989 [stationary sources, thousand tons]

	Total Emissions		Solids	SO_2
	[1,000 Tons]	[%]		
Estonia	576	100.0	262	210
Narva	312	54.2	162	125
Kunda	64	11.1	56	6
Kohtla-Jarve District	50	8.7	12	21
Tallinn	39	6.8	6	21
Latvia	159	100.0	38	59
Riga	30	18.9	–	–
Liepaja	15	9.4	–	–
Daugavpils	9	5.7	–	–
Ventspils	9	5.7	–	–
Lithuania	424	100.0	41	188
Trakai District	83	19.6	–	–
Akmene District	76	17.9	–	–
Mazeikiai District	67	15.8	–	–
Vilnius	32	7.5	–	–
Kaunas	26	6.1	–	–

Source: Central Statistical Bureau of Latvia (1998)

6.5 Baltic States and Security

The regional cooperation of Baltic States depends upon the following foremost factors:

- Cooperation between the three countries;
- Nordic states and neighbouring countries;

- All other Baltic countries like Germany and Poland;
- Membership of the three states in the European Union;
- Membership of the states in NATO and the security issues.

Let us now discuss the EU and NATO in a little more detail. The expansion issue is of primary importance. Since the integration is deeper and the countries concerned are of different types as compared to other integration. According to the Baltic Seas Treaty, EU can grow by Central and Eastern European countries, each having their own jurisdiction. Larger member states will give up their second Commissioners. However, a system needs to be devised by EU so that the big members are compensated by proportional weights in their voting rights. The cooperation of the commission is also a crucial factor which will inevitably lead to some disagreements. This accounts for Finland's hesitation to raise this question in the meeting. The changes in agricultural policies and financing are also equally vital issues.

In spite of these problems, the EU is fully committed to expansion. It will add millions new members as well as giving impetus to the economic reforms and structural changes in the Baltic States. This will also affect Russia and Ukraine. In the Helsinki Summit, Estonia, Latvia and Lithuania were officially invited to start negotiations. According to public officials, Latvia may be capable of completing these negotiations by 2003. Perhaps they will become a full member of the Union by 2005. Since present economic conditions in Lithuania are not promising and the process of economic reform is going slowly, the invitation is a promissory note which must be paid in the future by restructuring the economy. Latvia is making important steps towards such reform in the areas of education, agriculture, labour, health care.

Lithuania has to drastically reduce the state subsidy. After Estonia was invited before the other two countries, irritation developed between the three states. The situation is normalising now and the three countries realise that their secure and successful future lies in mutual cooperation and not in selfish isolation. There is a concern that the negotiation of the second round countries [namely Latvia and Lithuania] may slow down the negotiation with Estonia and there is fear that its membership will be delayed or even postponed. This is the reason why Estonia is trying to portray itself as a Nordic country and group Lithuania with Poland and Central European Counties and is placing Latvia somewhere in between.

Logically, Estonia should stress its simultaneous membership in both groups. EU expansion is based on political criteria. Although Russia had a positive attitude toward Finland with EU they are not in favour of Baltic integration unless the alleged harassment of the Russian minority is effectively removed. They would rather divide the relationship with EU on MFN [most favoured nation] principles rather than with the Baltic States. However, it is Russia's intent to keep a good relationship with EU and it is not expected that it will push this point to the extreme. It is estimated that when Baltic States become members, about half of

Russia's trade will be with the EU. After developing closer economic ties with Russia and deepening integration with EU, the Baltic States will make connections with the East as well as the West.

Although Baltic cooperation stated long before Benelux and Nordic experiments, there was no cooperation in security or free trade in agriculture. After independence and a different pace of economic reform in these countries, they are still coping with agricultural issues. However, the old issues still exist in the EU and there is still quarrelling with respect to genetically modified food as well as the banana war between USA and EU. The fishing rights quarrels between Estonia and Latvia have now been solved. There are many instances, including the Baltic Free Trade Agreement, which have helped both countries develop their markets. After these countries join the EU, regional cooperation will be more intense. Lithuania has now a bigger market for their export goods. Russia and Latvia will become the biggest trading partner for Lithuania, after Germany.

If a secure Baltic States region can develop as a member of EU, it can become a model for regions elsewhere in the world. Even if the EU decides to include the Baltic States one by one, it is in the interest of these countries to fully integrate since this experience will be useful at the time of integration with EU. The Baltic integration process is only possible due to the active support of Nordic countries. The attitude towards Baltic States has changed dramatically and these countries are buying arms both from NATO countries and Sweden. The development of the Baltic Defense College and other military and peace-keeping units are also contributing to NATO and European Defense and Security Identity [ESPI]. Joining NATO is much more complicated than joining the EU. Since countries like Finland and Sweden are not members of NATO, it also involves USA, Russia and EU. After the cold war, the Baltic Sea region has been evolving as the areas of Peace, Stability and Economic Growth. However, membership to EU and NATO should be total and not partial in order to maximise advantages and potential.

As previously mentioned, security remains a basic issue in the Baltic States. Although Russian influence is waning, old habits persist. It is important that the Baltic States become members of both EU and NATO. 1997 was an important year for Baltic security. EU decided to expand and invited Estonia to start accession negotiations on the fast track level. The NATO-Russian Act and Joint NATO-Russian Permanent Council have some direct relevance to Baltic security. On January 1, 1997, the Baltic States implemented their Free Trade Agreement by lifting trade barriers on agricultural goods.

An institutional framework of cooperation and the removal of tariff barriers enable the Baltic States to move toward four freedoms: freedom of movement of people, capital, goods and services. Russia opposes Baltic NATO membership while USA and Denmark at best offer lukewarm support. However, Baltic insecurity is also the same as European insecurity. The Baltic States are practical testing grounds. NATO membership is essential to ensure secure independence. It is quite possible that Russia can use the minorities in the states to justify acts of aggression. Although countries like Finland support the inclusion of Baltic States in EU, it thinks that there is no reason to sink the membership project of all Baltic States by inviting all three Baltic Sea States.

Such a move is impossible without close cooperation between the Baltic States. The enormous power anomalies between the three Baltic States determine the need for Baltic integration in NATO. A demonstration of massive force is necessary to stop another tragedy like Kosovo. Strong buffer states can serve Russian interests better than weaker states. Baltic States pay hard currency for Russian gas. In the past, heterogeneous security interests needed limited cooperation. However, the events of the past few years imply that more cooperation is needed for security. This is the reason why countries like US, Belarus and Ukraine are all involved in Baltic Sea projects. Enlargement of EU is the most single event to stabilise and increase the security. Accession implies both security benefits and greater obligations.

Developing an institutional structure of cooperation between Baltic Sea countries will promote reciprocal activity. EU's role will be a crucially unifying factor, particularly in respect to financing. As Russia becomes more stable and flexible, it will become a beneficial force for Baltic countries. A stable Russia integrated with Europe is beneficial to all countries on the continent. NATO's enlargement may not have a substantial integrating effect since countries in the region have different perceptions of security. The participation of Baltic States will provide variety, as well as deepening and broadening forms of cooperation. The Baltic States must catch up to the countries in Central Europe in order to minimise asymmetry in economic, political and military spheres.

References

Albach H. (1992): The Tranformation of Firms and Workers, Working Paper, Uppsala University, Uppsala

Blackburn N., Hagstrom A., Wikner J., Cuadros-Hansson first name missing, Bjornsen R. and Koefoed P. (1998): Rapid Determination of Bacterial Abundance, Biovolume, Morphology, and Growth by Neutral Network-Based Image Analysis, Paper Presented at the Conference *Applied and Environmental Science & Technology*, Easton, June 15, 1998

Braband J.M. van (1996): *The Political Economy of Transition*, Routledge, London, New York

Canfield J.L: (1993): The Independent Baltic States: Maritime Law and Resources Management Implications, *Ocean Development and International Law* 24 (1), pp. 1-39

Central Statistical Bureau of Latvia (1998): *Foreign Trade of Latvia,* Riga

Chepesiuk R. (1997): A Sea of Trouble, *Bulletin of the Atomic Scientists* 53, 40

Dellenbrant J.A. and Norgaard O. (1994): The Politics of Transition in the Baltic States, Department of Political Science, University of Umea, Department of Political Science

Estonian Nature Management Scientific Information Centre (1990): *Environment '89* Tallinn

Fischer M.M. and Getis A. (1999): New Advances in Spatial Interaction Theory, *Papers in Regional Science* 78, 117-118

Fischer M.M., Hlavackova-Schindler K. and Reismann M. (1999): A Global Search Procedure for Parameter Estimation in Neural Spatial Interaction Modelling, *Papers in Regional Science* 78, 119-134

Goskomstat (1990): *The Economy of the USSR; Ekonomicheskoye i Sotsialnoye Rarvitive Soyuznich Republik*

Government of Latvia (1990): *Latvian Environmental Statistics,* Riga

Government of Latvia (1994): *Monthly Bulletin of Latvian Statistics*, Dec. 1994, Riga

Government of Latvia (1994a): *Economic and Social Development of Lithuania*, Dec. 1994, Riga

Government of Latvia (1995): *Monthly Bulletin of Latvian Statistics*, Feb. 1995,

Government of Latvia (1995a): *Economic and Social Development of Lithuania*, Jan. 1995, Riga

Government of Lithuania (1990): *Lithuanian Environmental Statistics,* Vilnius

Haavisto T. (1997): *The Transition to a Market Economy*, Edward Elgar, Cheltenham

Hroch M. and Klusakova L. (1995): *Criteria and Indicators of Backwardness*, Charles University, Prague

Institut für Meereskunde (ed.) (1988): The Baltic Sea Environment. History – Eutrophication – Recruitment – Ecotoxicology, *Kieler Meeresforschungen* [Special Edition No. 6], Kiel

Karlsson M. and Arkadie B. van (1992): *Economic Survey of the Baltic States*, New York University Press, New York

Land T.O. (1997): Berlin: Capital on the Move, *Contempory Review* 270, 68

Laurin F. (1991): Scandinavia's Underwater Time Bomb. As Their Containers Rust Away, Chemical Weapons – Dumped Hastily in the Seat at the End of World War II – Have Been Showing Up in Fishermen's Nets, *Bulletin of the Atomic Scientists* 47 (2), 10

Lejins A. (1999): *Baltic Security Prospects at the Turn of the 21st Century*, Kikora Publishers, Helsinki

Ozsoy E. and Mikaelyan A. (1995): Sensitivity to Change: Black Sea, Baltic Sea and North Sea, Kluwer, Boston

Raike A. and Pietilainen O.P. (1998): Baltic Sea. An Overloaded System, *WWF Baltic Bulletin* 2/98, 12-18

Storn R. and Price K. (1997): Differential Evolution – A Simple and Efficient Heuristic for Global Optimisation over Continuous Spaces, *Journal of Global Optimization* 11, 341-359

Strandberg B., Bavel B. van, Bergqvist P.A. and Broman D. (1998): Occurrence, Sedomentation, and Spatial Variations of Oranoclorine Contaminants in Setting Participate Matter and Sediments in the Northern Part of the Baltic Sea, Paper Presented at the Conference *Applied and Environmental Science & Technology*, Easton, June 15

Toro T. (1991): Baltic Bombs Surface in Fishing Nets, *New Scientist* 131, 9

Voipio A. (1981): *The Baltic Sea,* Helsinki

World Bank (1989): *Country Statistics 1989*

World Bank (1991): *Country Statistics 1991*

Part C: Mobility in Transport Systems in Europe

7 The Role of Infrastructure for Expansion and Integration

Roger Vickerman
University of Kent at Canterbury, Department of Economics, Keynes College

7.1 Introduction

Infrastructure is seen as playing a key role in the process of integration. Since infrastructure has typically been seen as a public sector responsibility, in its planning if not also in its finance, it has tended to reflect the priorities of the state. In a Europe of separate nation states this has implied the absence of a Europe wide set of infrastructure plans which has often been argued to lie at the heart of Europe's increasing lack of competitiveness in world markets. This arises in two ways. First there is the lack of physical connections at an international level, such that the equivalent of the US Interstate Highway network was not conceived until the 1990s. Secondly the management of the infrastructure often causes even more difficult problems: the lack of consistency in electrification and loading gauges in the railways and the problems of diverse air traffic control authorities are good examples.

In this contribution therefore we outline the role of transport infrastructure in economic development in the specific context of the European Union, we consider ways of modelling and measuring this relationship and the development of policy. In particular we examine the problem of combining policies which seek to improve the competitiveness of the EU and those which seek to increase cohesion and promote convergence of GDP per capita, especially in the context of expansion of the EU to include Central and Eastern European countries.

7.2 Treaty and Statutory Basis of EU Action

As a starting point it is important to establish the role of the EU in the transport domain, and the possible scope for developing this role. The European Union it should be remembered is a treaty based organisation of independent states. The EU has power to act in those cases where supra-national activity is more efficient than national activity, and/or where there are important externalities or spillovers. All EU action is governed by the main treaties of Rome (1957), Maastricht (1992) and Amsterdam (1997). Treaties have to be negotiated and then agreed and ratified unanimously by all members. From the Rome Treaty, transport has been identified as one of those areas where the EU has jurisdiction, indeed it is one of only three areas [the others being agriculture and external trade] where a common policy is to be developed. The treaties do not, however, determine the precise nature or coverage of such policy. This is left to the routine policy making structure of the EU.

Policy decisions taken by the EU fall into three main types governed by the need for agreement between the sovereign members [it is actually much more complicated than this, depending also on who can take the decisions]:

- the most important decisions [e.g. those concerning a revision of the treaties, admission of new members] require unanimity;
- the most frequent decisions concerning policy are decided by 'qualified majority' [i.e. countries are weighted approximately by size, and decisions require a minimum numbers of weighted votes to be approved] – most decisions on transport are taken by this route:
- relatively minor decisions are taken by simple majority.

These decisions are taken variously by three broad institutional groups:

- The Council of Ministers is the body representing the 15 national governments – formally it takes most decisions and acts as the formal EU government.
- The European Commission is the body which constitutes the executive arm of the EU, it comprises 20 Commissioners [two from each of the five largest states and one from each of the rest], it proposes and implements legislation and polices it.
- The European Parliament is the directly elected representative arm of the EU, it responds to, and comments on, legislation and has to approve certain elements [such as membership of the Commission and the EU budget], but does not otherwise act like a normal legislature. An increasing number of matters is resolved by the co-determination procedure of Council/Commission and Parliament.

The European Court rules on disputes concerning interpretation of the treaties. Note that European law takes precedence over national law, and where there is any conflict on a matter which has been properly decided national law has to be changed to accommodate the provision of the European law. This is important in the case of transport since the European Court has had to make a number of key rulings on the scope of decision making and the lack of progress in developing the policy towards transport.

7.3 The Role of the EU in Transport and Transport Infrastructure

As noted above, the development of a common policy for transport is a treaty obligation dating back to the Treaty of Rome. This recognised from the outset the economic importance of transport in binding together the Common Market. However, very little was achieved for the first 30 years in terms of what could genuinely be called a common transport policy.

The main provisions were to simplify cross-border operations, principally by road, where the operation of vehicles registered in one member state would need to operate in another. These provisions were mainly under three heads: social, technical and fiscal. The social provisions related principally to regulations on drivers' hours. It was important to ensure that vehicles could be operated safely throughout the EU, and this also ensured that lower standards applying in one member state did not give its operators an unfair market advantage. Technical provisions had a similar purpose, it would be against the basic principles of a common market if the vehicles of one member state could not operate legally in another, but to achieve this matters relating to such factors as vehicle dimensions, weights etc had to be resolved. Fiscal matters relate to the taxation of vehicles and fuel. There was no move to harmonise tax rates, but vehicles would be allowed to cross borders without specific penalties. This provided for free access to all highways [except those subjected to a specific toll payable by all users] and for vehicles to cross borders with full tanks of fuel.

Little technical was achieved with railways, indeed, as the development of largely incompatible high speed rail technologies in the 1980s demonstrated, separate development for national industrial reasons continued in new as well as old technologies. The railway sector was burdened with the history of different loading gauges, different electrification systems and, most significantly, different signalling and train protection systems, all of which combined to make through operations without a change of locomotive difficult.

During this period little real progress was achieved in creating a genuine common market in transport services. For most of the period restrictive bi-lateral quotas governed cross-border truck operation and there were no fifth freedom or cabotage rights in which the vehicles of an operator in one member state could move loads between two other member states or within another member state. This led to grossly inefficient operations with a high proportion of underutilised or empty vehicles, raising operating costs and imposing external costs of congestion and pollution through increased vehicle movements. The gradual expansion of 'community quotas' helped to reduce this problem, but the burden of regulation continued to bear heavily on the road sectors due to restrictive national regulations designed to protect inefficient state owned railways.

On the railways there was little progress towards achieving 'inter-operability' in rail services [especially freight]. As well as the technical problems already mentioned, which often require a change of locomotive at frontiers, the difference in operating practices meant that even where expensive multi-voltage locomotives were provided, crews were not licensed to operate them outside their home territory so that international travel by train involved lengthy stops at border crossings, and for freight this increased the chance of unreliable or insecure delivery.

The development of common policies towards the finance and provision of infrastructure was also extremely slow. Much of the effort went on defining common accounting practices, especially in the rail sector, and mainly to avoid unfair discrimination between sectors rather than to develop an efficient transport infrastructure.

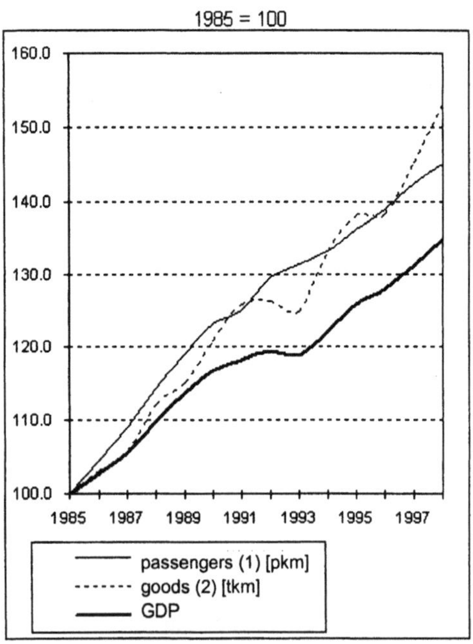

Notes: (1) passenger cars, buses+coaches, tram+metro, railways, air
(2) road, rail, inland waterways pipelines, sea [intra-EU]

Source: *EU Transport in Figures* (2000), Directorate General Energy and Transport, European Commission

Fig. 7.1 Transport growth EU15 [1985=100]

The problem which policy had to address had been growing during this period. Four main issues can be identified: *First,* the overall rate of traffic growth had begun to diverge from that of GDP growth, especially for freight traffic. This tendency increased as the pressure of globalisation and the completion of the single market, coupled with a long term real reduction in the price of transport encouraged longer journeys and freight hauls Fig. 7.1). *Secondly*, this growth was largely concentrated in one mode, road transport (although the fastest rate of growth was in air transport) (Fig. 7.2). *Thirdly,* this was associated with a continuing rise in the level of motorisation in the EU. *Finally,* despite this continuing growth there were considerable variations in the transport intensity of different economies, measured as the amount of transport per unit of GDP (Fig. 7.3 and Fig. 7.4). These four factors led to increasing pressures of congestion and environmental impact.

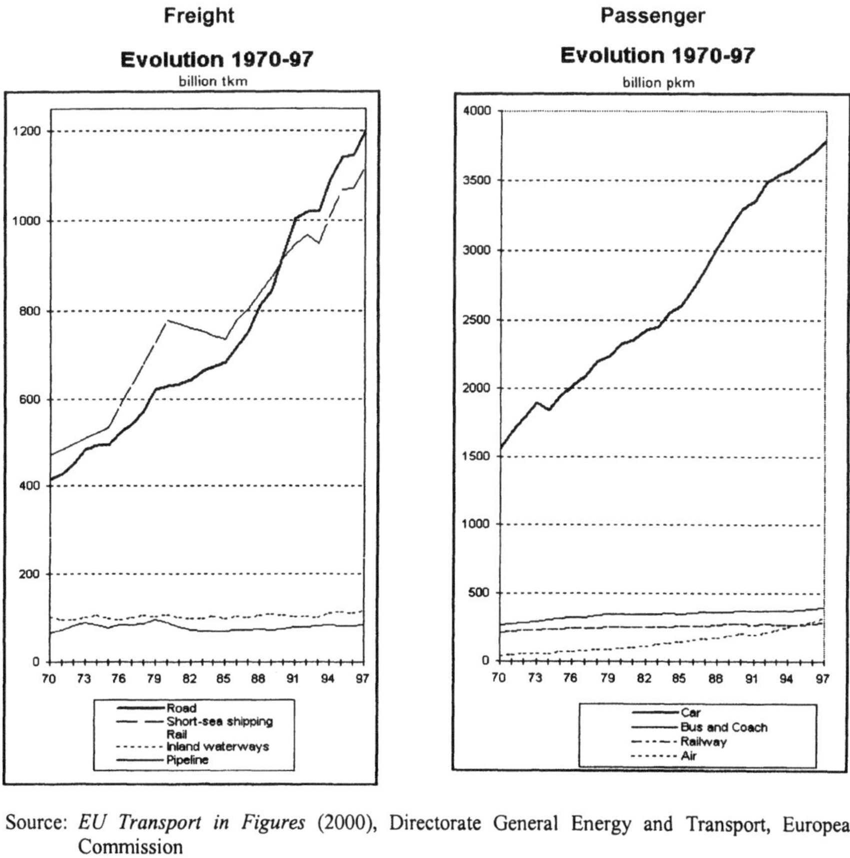

Source: *EU Transport in Figures* (2000), Directorate General Energy and Transport, European Commission

Fig. 7.2 Evolution of traffic by mode

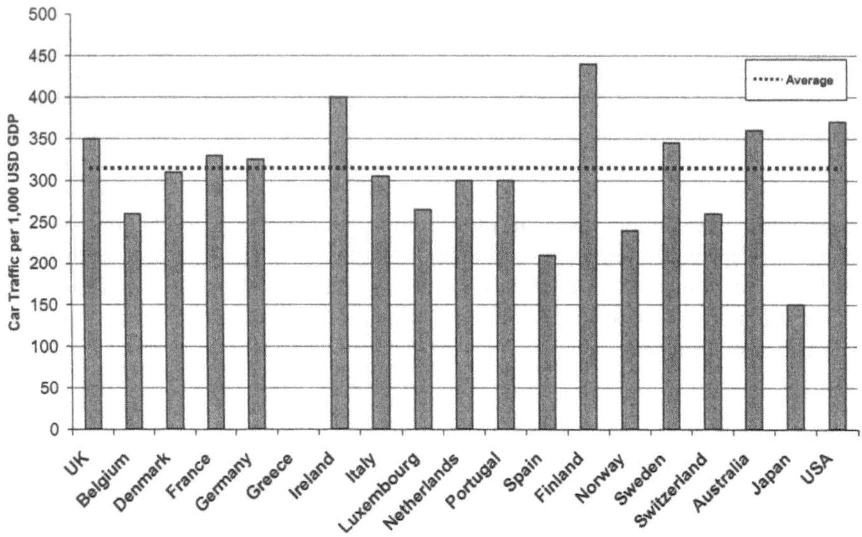

Fig. 7.3 Car traffic per 1,000 USD GDP

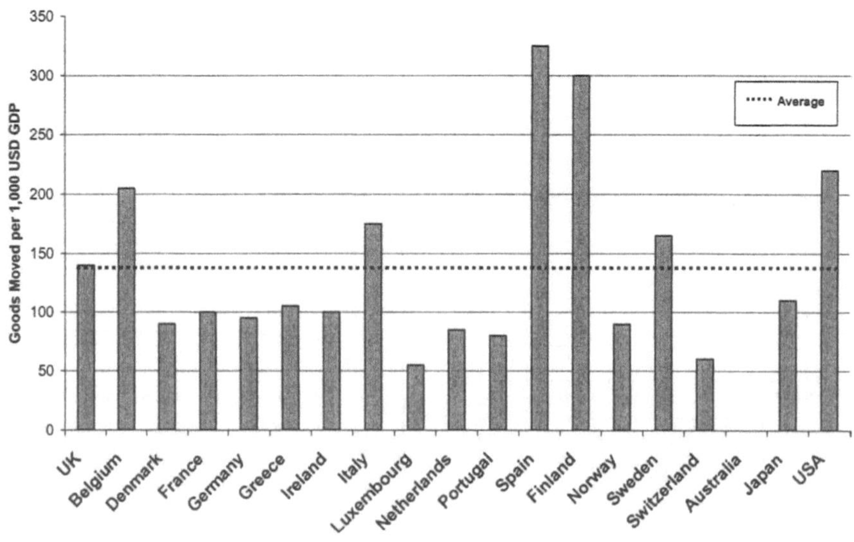

Fig. 7.4 Goods moved per 1,000 USD GDP

These pressures increased concern about lack of progress in the formulation of an effective common transport policy. This changed with some landmark European Court decisions in the late 1980s and early 1990s. *First*, the 'Nouvelles Frontières' case in 1986 reaffirmed the application of the 'common market rules' to air transport. Air transport had in fact only been added to the scope of the EU in the 1970s, most air transport being seen as a bilateral responsibility of the origin and destination countries, thus again preventing the utilisation of fifth freedom rights and other benefits of deregulation which had been achieved earlier in the US. *Secondly,* and more significantly, the European Court ruled, after a reference by the European Parliament, that the Commission had failed to meet the Treaty obligations in respect of transport. This led to a flurry of activity in the early 1990s.

The White Paper on *The Future of the Common Transport Policy* in 1992 made two main contributions. It set out the need for greater coherence in transport policy at the EU level, particularly with regard to such issues as sustainability, safety, inter-modality and inter-operability. In addition, it set the scene towards a common approach to pricing. This was not a return to the fixing of prices, which was attempted for road haulage in 1960s, but of ensuring fair competition between modes and the compatibility of national systems. Much of this was driven by a need to find a common approach to the external costs of transport which could ensure the consistent development of transport in the EU.

The White Paper's aims were supported further by the Maastricht Treaty on European Union and the 1994 White Paper on *Growth, Competitiveness and Employment.* These two documents identified transport infrastructure in particular as an instrument of integration. The designation of Trans European Networks [TENs] together with new financial instruments to ensure their rapid development, would help to achieve the twin goals of increased competitiveness [lower transport costs enabling the exploitation of scale economies] and cohesion [provision of better accessibility in lagging peripheral regions would unlock their development potential]. This was less an approach to the development of the transport sector than using transport as an instrument of macroeconomic policy through job creation to tackle unemployment. It appeared to resurrect a rather old fashioned Keynesian public works solution to unemployment and arguably tried to make transport policy do too much. This was particularly so in the apparent obsession with infrastructure rather than a broader approach to ensuring greater efficiency in the transport sector as a whole. We return to the details of this policy intervention in the following section.

Following this attention turned to the question of the external costs of transport and to the most appropriate way of charging for infrastructure use in a set of white papers relating to 'fair and efficient pricing'. These recommend following the basic principle of marginal social cost pricing following the definition of basic sets of 'prices' for each element of social cost. Such an approach addresses environmental issues, and some other elements of external cost such as accidents, effectively, but short-run congestion pricing may lead to problems of financing infrastructure in the longer term.

Finally to redress some of the emphasis given to high level infrastructure in the TENs programme the Commission turned to the concept of The Citizens' Network. This was more to do with providing access at the most local level and addressed the question of social inclusion at the individual level [in contrast to the wider regional level issue of cohesion].

7.4 TENs, TINA and Infrastructure Pricing

The underlying motive for the development of the TENs was, as discussed above, the need for a more coherent macroeconomic policy. This embodied the twin objectives of increasing both competitiveness and cohesion. Although there was a logical role for transport infrastructure in increasing the integration of the EU economy to achieve both aims, there was a danger of incompatibility between the centralising tendencies in the competitiveness objective [based essentially on the achievement of scale economies and thus potentially increased spatial concentration of economic activity] and greater cohesion. This problem is discussed more fully in the next chapter.

Trans-European Networks [TENs] were defined for each of the following main modes:

- high speed rail,
- conventional rail,
- road,
- airports,
- seaports,
- inland waterway,
- pipelines,
- telecommunications.

A widely based network is defined for each mode, but these depended a lot on bottom up lobbying by regions. This meant that the different networks looks remarkably similar, few strategic decisions were taken at the EU level to define the most appropriate network for each mode. In addition the priorities for key elements were established on the pragmatic grounds of what could be done quickly.

This led to the identification of 14 key projects at the Essen Council meeting (Table 7.1). These cover most of the EU [every member state is involved to some extent] and most modes, but there is a clear emphasis on rail. The selected projects were based to a large extent on these projects' maturity and their ability to be financed. They were to serve to a large extent as demonstration projects rather than being the key projects which were needs to achieve the main goals [except to the extent that the projects already well developed and able to be financed were more likely to serve these aims]. These projects were not financed directly by the EU [except for certain studies and some very basic support] such finance would depend on member states, regions and the private sector. Direct EU funding was expected to contribute less than 5 percent of the total TENs investment cost of €400bn and the Essen projects' total cost of €110bn.

There were problems of reaching agreement on a proposed new Community financial instrument, and it was assumed that, following experiments such as the Channel Tunnel in the 1980s, private finance would contribute a substantial share. Euro bonds and a European Infrastructure Fund managed by the European Investment Bank made a contribution to the €110bn total cost [of which just €13bn was spent in the five year period 1993-1998]. Direct EU finance contributed €3bn in this period; it was proposed that this should increase to €5bn in the period 2000-6. As will be seen from Table 7.1 many of these key projects will not be complete until well after the end of this period.

Table 7.1 Essen list of priority TENs

Project	Type	Member State	Completion
HST North-South	High speed rail/multi-modal	I/A/D	2005+?
HST North [PBKAL]	High speed rail	B/UK/NL/D	2005
HST South	High speed rail	E/F	2005?
HST East	High speed rail	F/D/L	2005?
Betuwe Line	Rail/combined transport	NL/D	2005
HST Lyon-Trieste	High speed rail/multi-modal	F/I	2005+?
Greek Motorways	Road	GR	2005
Lisbon-Valladolid	Road/multi-modal	E/P	2005+
Cork-Dublin-Belfast-Larne	Rail	IRL/UK	2000
Malpensa Airport, Milan	Air	I	2000-
Øresund Fixed Link	Rail/road	DK/S	2000
Nordic Triangle	Multi-modal	(N)/S/SF	2005+
West Coast Main Line	Rail	UK	2005
Benelux-Ireland Road Link	Road	B/UK/IRL	2005+

Note: PBKAL is the Paris-Brussel-Köln-Amsterdam-London HST network.

The Essen list was concerned solely with the EU's internal transport needs. It was rapidly recognised that there was a crucial need to consider the extension of the European network to the countries of Central and Eastern European Countries [CEEC] in transition. These were poorly developed networks, in a bad state of repair, and badly integrated with the EU. They were needed both to help the CEEC countries in their candidature for EU entry, but also to help the EU develop its trade with these countries, especially given the amount of direct investment being placed in these countries. A transport infrastructure needs assessment [TINA] was carried out for these areas. This produced a set of key corridors, as opposed to the individual links of route networks, for development. Ten Pan

European corridors and four Pan European transport areas were defined including 20,000 km of rail, 18,000 km of road, plus air, sea and river ports, with a total cost of €90bn. These corridors did provide a more focussed top-down network approach to the problem, but often had more to do with current EU's needs for better transport in CEEC than the CEEC's own needs, and there was relatively little assessment of specific needs of the economies of the CEEC.

7.5 Assessment of Policy Objectives and Achievements

The main conclusion to be drawn from this review is one of a general confusion of policy objectives in EU transport which is particularly acute in the case of policy towards infrastructure. Three key problems emerge here. *First* there is a conflict between policies towards achieving internal consistency in transport sector and the role of transport in wider economy. *Secondly,* as we have already identified, there is a potential conflict between competitiveness and cohesion. *Thirdly,* there has been a possible excess emphasis on large prestige projects of EU interest to the exclusion of local projects which could have wider significance, e.g. the emphasis on high speed rail

A number of more detailed problems have also emerged. There has been slow progress in completing the single market across all modes and a failure to develop multi-modal solutions to transport problems. There has been an emphasis on infrastructure without development of adequate analysis of specific impacts, and an emphasis on new infrastructure to the exclusion of better management of existing infrastructure. Although there has been a welcome increased emphasis on sustainability, this has generally been in the context of accommodating increasing mobility as an indicator of welfare enhancement, rather than a full evaluation of the real needs for transport in the economy to ensure both environmental and economic sustainability. The emphasis on fair and efficient pricing is also a welcome advance in achieving a more consistent set of policies, but this has been undertaken without adequate analysis of implications for funding.

There two areas where the EU has failed to provide an adequate framework. First there has been a lack of direct financial provision. The transport budget has been relatively modest and thus much infrastructure funding has been provided through the Structural and Cohesion Funds without a full transport assessment. Generally the assumption here has been that because lagging regions lack infrastructure, the provision of more must be a good thing, an assumption we shall challenge in the following chapter. There has also been the lack of a common appraisal framework for infrastructure, both between modes and between countries.

Secondly, there is the absence of a single championing authority. This has led to reliance on local or national studies and made the policy subject to the action of competing interest groups in the European Commission. This has led to a failure to take a strategic view of EU needs as opposed to regional and national interests.

References

Banister D., Capello R. and Nijkamp P. (eds.) (1995): *European Transport and Communications Networks: Policy, Evolution and Change*, Wiley, London

European Commission, Directorate General Transport (1998): *Integrated Strategic Infrastructure Networks in Europe,* Final Report COST 328, Office for Official Publications, Luxembourg

Turro M. (1999): *Going Trans-European: Planning and Financing Transport Networks for Europe*, Pergamon, Oxford

Vickerman R.W. (1995a): The Regional Impacts of Trans-European Networks, *The Annals of Regional Science* 29, 237-254

Vickerman R.W. (1995b): Location, Accessibility and Regional Development: The Appraisal of Trans-European Networks, *Transport Policy* 2, 225-234

Vickerman R.W. (1997): High Speed Rail in Europe – Experience and Issues for Future Development, *The Annals of Regional Science* 31, 21-38

Vickerman R.W. (1998): Transport, Communications and European Integration. In: Pinder D. (ed.) *The New Europe: Society and Environment*, Wiley, London, pp. 223-238

Vickerman R.W., Spiekermann K. and Wegener M. (1999): Accessibility and Regional Development in Europe, *Regional Studies* 33, 1-15

8 Restructuring of Transportation Networks

Roger Vickerman
University of Kent at Canterbury, Department of Economics, Keynes College

8.1 Introduction

The discussion in the previous chapter has shown that our understanding of the ways in which transport infrastructure affects economic development and the process of integration is still seriously limited. The major problem we face is that there is an assumption that infrastructure must have a positive effect on growth, and hence on both competitiveness and cohesion. In this contribution we shall review the possible theoretical approaches which will enable us to be more confident in our ways of appraising new infrastructure projects.

Why do we need theory? If we lack a full theoretical framework it will be extremely easy to make erroneous assumptions about the way infrastructure works in the economy. However, there are several different approaches which have taken a very different philosophical stance. Economics models have for the most part ignored the spatial dimension. Infrastructure is seen as an input to the productive process, enabling the growth of output, just as any other input. Geography models have essentially taken the reverse position, they have concentrated on the effect of infrastructure on accessibility, but thus ignored the way in which the changing price of accessibility might lead to substitution between inputs. Transport models [and, in particular land-use transport models] have had a less theoretical basis, but have integrated the accessibility of destinations with decisions about transport more fully. However, such models have largely ignored the overall impact on the activities for which the transport is undertaken. More recently, so-called 'new economic geography' models have offered a more formal way of combining the spatial element into a consistent economic framework. We consider some of the main contributions in turn in the following sections.

However, acceptable the theoretical basis of a model is, it needs to be capable of being applied in a real world situation. This requires careful consideration of both issues of measurement and methods of appraisal.

8.2 Review of Basic Theoretical Approaches

The basic role of infrastructure in the economy can be seen as twofold. Infrastructure as an input can be seen as a public good. Better infrastructure is neutral in its treatment of different sectors, but enhances the effective productivity of all other inputs. However, as an input, infrastructure can be traded off against other inputs such as private capital, labour and land markets. Although infrastructure might be seen to be broadly complementary to private capital, raising the marginal productivity, in the case of labour and land markets it may be an effective substitute. Thus, better accessibility enables firms to widen labour market areas with consequent effects on both local wage levels and employment.

Traditional approaches to infrastructure and its pricing have suggested that most infrastructure, because of its public good characteristics, should be provided by the public sector and be free at the point of use. More recently this view has been challenged, suggesting that the impact of public infrastructure on private productivity is that it raises the rate of return on private assets, which can be used to pay for the infrastructure provision. This view tends to regard public infrastructure as more of a club good, in which even if there is no direct payment at the point or time of use, it is possible to exclude non-payers from use and therefore a regulated private sector provision is feasible.

If this is true then most of the benefits of infrastructure should be able to be captured through an appropriate charging regime. If infrastructure promotes local or regional or national economic growth, then this implies increased benefits [profits] to users of that infrastructure for which they should be prepared to pay. In a world of perfect competition amongst users, they will be prepared to pay an amount [their marginal willingness to pay] exactly equal to the value of the benefits they derive. This depends on the critical assumption of perfect competition, any departure from this means that the marginal willingness to pay is not exactly equal to the benefits of the infrastructure.

However, assuming perfect competition for the moment, any attempt to identify wider economic benefits above those directly attributable to users leads to a danger of double counting. For example, increases in accessibility are often associated with increases in land rents, but these are simply a transfer of user benefits to factor owners and are not additional to the user benefits. As a general rule therefore, it has been assumed that no wider economic benefits should be identified in transport appraisal.

There is an additional worry where public funds are used to finance infrastructure, whether these are borrowed or raised from taxation. The use of public funds diverts finance from private investment by raising the cost of private finance through higher interest rates or higher taxation. This crowding-out of private investment by public investment does, however, assume that the public infrastructure is less productive than the private capital, but the power of governments to underwrite borrowing or raise taxation enables it to divert investment to its schemes. This would lead to investment in public infrastructure reducing the potential rate of growth of the economy.

Hence a key research question becomes the extent to which the productivity of public infrastructure investment is greater or smaller than that of private capital. If it is greater then this would outweigh the crowding out effect. If it is greater then this would confirm the public good nature of infrastructure, and this would suggest that, since users of public goods tend to undervalue their willingness to pay of such goods, that there may be a prima facie case that there are wider benefits which are not captured by user charges. In such a case a private sector provider may find it difficult to capture any wider benefits.

8.3 Aggregate Productivity Models: The Aschauer Debate

The question of the productivity of public infrastructure investment was addressed by Aschauer in a set of papers in the late 1980s. Aschauer wanted to demonstrate that the cut back in public investment in the US had been damaging to economic growth. Aschauer used an aggregate production function which introduced infrastructure investment [I] as well as the conventional labour [L] and private capital [K] inputs in a Cobb-Douglas form:

$$Y = f(L, K, I) = L^{\alpha} K^{\beta} I^{\gamma} \text{ (or } L^{\alpha} K^{1-\alpha} I^{\gamma}\text{)}$$

The alternative form attempted to preserve the usual assumption of constant returns to scale of the conventional private inputs of labour [L] and capital [K] in a competitive economy. Using both time-series and cross-section data Aschauer (1989) found evidence consistent with infrastructure having a substantial impact on the rate of growth of GDP [Y]. US data suggested an output elasticity of public infrastructure much higher than previously thought [up to 0.45 or 0.5] and thus the implicit rate of return to such investment was very high, well over 100 percent. In such a situation the rate of return on public infrastructure relative to that on private investment would outweigh the problem of 'crowding out', at least in the long run.

Despite this strong result there was considerable scepticism about the reliability of such a finding. Attempts to replicate the results found some consistent evidence both in other countries and across countries and regions, although the output elasticities were usually smaller, but there were also contradictory results. The rate of return to soft infrastructure such as education was often higher than that to physical infrastructure, and that to transport was often amongst the lowest. There were considerable problems of misspecification and econometric error identified in the original work.

One of the main difficulties is understanding the mechanism which operates at an aggregate level. Infrastructure is used in different ways in different sectors, and this is totally lost in the aggregate productivity model. There is thus a need for more disaggregate/microeconomic studies of the role of infrastructure.

8.4 Input-Output Models

One way of achieving a degree of disaggregation is through the use of Input-Output Models. These detail linkages between sectors including their use of transport and can detail the origins and destinations of goods traded. This enables the possible tracing of impacts of an improvement in transport through the sectors which really use transport, e.g. those with high transport costs or those with a high elasticity of demand. However, input-output models use fixed 'Leontief' production coefficients, which makes it difficult to allow for substitution between transport and other inputs, especially in a dynamic context in which if transport infrastructure does have an impact on growth we would expect to see these coefficients change. Input-output data is often poor and fails to account adequately for in-house transport, business transport by personnel etc. Local and regional input-output data is particularly poor.

Some use of input-output models has been made in detailed studies, particularly where construction impact is large, e.g. airport expansion study in the UK, and new infrastructure links in Denmark.

8.5 New Economic Geography Models

The epithet 'new economic geography' was given by Paul Krugman to an important strand of work which has helped to bring space back into a formal economic model structure using some recent developments in a number of areas of economics, including consumer theory and including new trade theory. This approach emphasises the role of scale economies and transport costs and how these two trade-off against each other. Scale economies, which imply larger market areas for firms, will exist where firms face imperfectly competitive markets. Increasing transport costs, *ceteris paribus*, imply smaller market areas. Thus there can be a trade-off between scale economies and transport costs, but depending on both the relative rate of change and the initial starting values, the result can involve ambiguous responses to a given fall in transport costs.

This gives us a new insight into the way transport changes will have a regional impact. Regions can benefit from increased competition from outside which may follow a reduction in transport costs. This may lower the profits of firms in the region, but will increase consumers' welfare from lower prices. However, this may also lead to jobs moving out of a region – the argument which we can term 'the two-way road' and is analogous to the change in accessibility in core regions which can arise from an improvement of transport links to peripheral regions. Who benefits most will often depend on which firms can exploit their potential comparative advantage most quickly and effectively. These models also enable us to use a general equilibrium approach which allows us to examine, not just the direct impact of transport changes on initial users, but also the way in which changes in the transport sector may induce further changes in the transport-using sectors and in the links between them.

8.6 Measurement Issues

Accessibility and economic potential: The traditional way of measuring the changes brought about by investment in new transport infrastructure is through measures of accessibility. Accessibility is a measure of the spatial price of undertaking activities, since it refers to the cost of reaching [a set of] specific destinations. One of the problems with the use of accessibility as a measure of the cost of access is that it is also frequently interpreted as a measure of welfare in itself, improved accessibility is thus seen as a goal in its own right.

The traditional measure of accessibility is based on the gravity model. If the number of trips between i and j is given by $T_{ij} = a\, Y_i\, Y_j\, f(c_{ij})$, in which Y_i and Y_j are some measures of the weight of the origin and destination, and $f(c_{ij})$ is a function of the generalised costs of moving between i and j, then we can write the accessibility of i to j as $A_{ij} = a'\, Y_j\, f(c_{ij})$, where a' is a constant. The total accessibility [economic potential] of i is then $A_i = \Sigma_j A_{ij}$.

The main problems with this measure are the need to establish appropriate values for the zonal weight Y_i, e.g. population, income, output, retail expenditure and the need to establish parameters of generalised cost and assume they are unchanged with transport change. Such accessibility based measures are usually based on a single mode of transport, which may not give an adequate picture of the effective accessibility of a given location. Locations with different sectoral structures can be expected to have different transport needs. The assumption of continuity in the index is also a problem. This may be in order for purely road measures, but for public transport and rail it is necessary to allow for access to network; for high speed rail and air this is even more the case. We also need to determine how to deal with the relative attraction of the home and adjacent zones [the 'self-potential' problem] to ensure that we make some allowance for close by facilities without these having the effect of swamping the measure.

Some of these problems can be dealt with by modifying zone structures, e.g. by using a raster cell base, using mode split data to weight different modal accessibilities in order to produce a synthetic 'aggregate' accessibility, or by basing accessibility on a 'best mode' figure, arguing that effective accessibility is determined by the mode which offers the lowest generalised cost, but that this mode may differ from place to place.

Alternative approaches to the gravity model basis include time-space mapping, which translates all distances into times and maps these to give a visual impression of the accessibility surface, and daily accessibility, which measures the total population accessible within a given time frame [e.g. how many can be reached within five hours, which allows a return journey to be completed within a single day].

A number of conceptual problems remain, however. As already mentioned we have to resolve whether accessibility is just a price or whether it can be interpreted as a welfare indicator such that maximising accessibility becomes a goal in its own right. We also have the problem of relative or absolute accessibility. Is what is relevant simply the change in accessibility for a given location resulting from a change in transport provision, which can be measured just by ΔA_i or do we need to show how accessibility is changing relative to an overall measure of accessibility for all relevant places, such as $\Delta A_i/\Delta\Sigma_i A_i$ or even the change relative to that place with the best overall level of accessibility, i.e. $\Delta A_i/\Delta A_j^{max}$? Whichever measure is chosen we then have to resolve the question of the expected relationship between the change in accessibility, ΔA_i and the change in income or welfare, ΔY_i.

Measuring the contribution of infrastructure: Measuring infrastructure itself poses some serious problems for analysis. One of the main problems with the aggregate analyses discussed above is that they need to determine an appropriate measure. Studies have used both physical and monetary measures. Physical measures are reasonable in studies which are using a single mode, for example road infrastructure, since they can introduce quality measures such as the number of lanes. For an aggregate measure of all types of infrastructure in the economy as a whole a monetary measure is normally used. For public infrastructure the use of a monetary measure may be less appropriate than for private infrastructure. In the case of the latter it is reasonable to suppose that the price paid for investment goods is reasonably close to the value of those goods. However, the uniqueness of each piece of infrastructure investment, its sheer size and the likelihood that it is procured out of public funds all mean that the actual price paid is less likely to be to be a good indicator of economic worth.

A typical means of valuing capital stock is through the flow of investment, but again the lumpiness and infrequency of investment in large infrastructure projects in any one location makes the flow measure a less appropriate measure of the stock of infrastructure. This biases the measure of the value of infrastructure towards its construction impact rather than to its value in use which is determined by the stock. Furthermore the use of a monetary value of the infrastructure may not tend to be an accurate measure of the quality of infrastructure in use – investment cost will for example often reflect the engineering difficulty of construction, the need to protect environment etc., and not reflect the value of the infrastructure to the user, especially where there is no direct payment for use.

This leads us to the need to consider the role of infrastructure costs in transport costs and total logistics. It is important to separate out these two effects since the impact on transport may be only a short run phenomenon until firms can reorganise both their transport and total logistics to accommodate changing transport costs. Thus in the long run the location of production, of distribution centres and the production process itself can all be altered to allow for infrastructure changes. This is unlikely to be a continuous process in that such changes are often costly exercises, leading to inertia which results in discontinuities and threshold effects. In particular there may be important network scale economies which means that individual changes to a region's infrastructure do not have an impact until a whole network has been improved. At this point there may be a major step-change in the economic organisation and welfare of the region.

8.7 Evaluation and Appraisal

In this section we shall review briefly the main approaches to appraisal and the main implications of the previous discussion for improvements to this process.

***Ex ante** and **ex post** studies*: Most appraisal is carried out on an ex ante basis, as part of the investment decision process, and before and after studies to check in the accuracy of appraisal are rarely carried out. Given the size of investments being considered this is surprising, but there are major conceptual problems in achieving this. The major difficulty is one of achieving consistency and dealing with the counter-factual. Given the length of time over which such investments take effect it is often difficult to ensure measuring the same thing before and after and the results will be affected by extraneous changes. Moreover the serious problem of determining cause and effect, when a number of changes occur, recursively, over a long period, remains.

As well as the aggregate studies in the Aschauer tradition discussed above, a number of case studies of infrastructure networks have been carried out, for example of American railroads in the nineteenth century and more recently of the

UK roads programme and the EU TENs. The *ex post* studies of American railroads have produced mixed results, there is a suggestion that the social product in the American West may have been somewhat higher as a result of the railroads, but that it is unlikely that the railroads were the major cause of economic growth. This is rather like the finding of the Biehl study for the EU which suggested that although infrastructure could be an important aid to growth where the lack of infrastructure was the main constraint on such growth, but in other cases it was irrelevant, whilst those regions which were infrastructure rich were typically those which were better able to fund higher levels of infrastructure or needed to divert a higher proportion of their output into infrastructure to support the growth which originated elsewhere.

The UK roads and EU TENs studies were *ex ante* studies which have used large scale aggregate macro-modelling approaches. These tend to make fairly major simplifying assumptions which assume rather simplistic causal relationships between infrastructure improvement, productivity and economic growth and consequently produce very optimistic estimates of the likely impact of network improvements. Whilst it might be argued that there are important network effects which may lead major programmes to have a larger impact than individual projects – these networks need to be assessed in terms of their sectoral and spatial impacts and not just at the aggregate macroeconomic level.

Individual large projects have attracted considerable attention, in terms of both *ex ante* and *ex post* studies. A number of studies of the various French TGV lines have been completed showing a rather mixed set of effects on regional development. There has been a tendency towards the centralisation of economic activity around the main stations served, though not necessarily towards Paris at the expense of provincial cities. Attempts to generate economic activity through the creation of intermediate stations on lines have, however, not been very successful. More rapid communication between Paris and such cities as Lyon and Lille has been the catalyst for considerable reorganisation within firms.

In Europe the two other major projects which have received considerable attention are the Channel Tunnel and the various Danish bridge and tunnel schemes. These are all very large projects involving the completion of missing links in the network. Whilst these may be important in traffic terms, the evidence from both *ex ante* and early *ex post* studies is that they may be less important in regional economic activity terms, although there are considerable variations here between schemes which serve as major international links such as the Channel Tunnel [and the proposed Fehmarn link between Germany and Denmark] and those which serve interregional needs within a country such as the Great Belt link. The Oresund link, just opened in July 2000 is interesting in that although international in character it has a primarily local function linking the two neighbouring cities of Copenhagen [Denmark] and Malmo [Sweden].

Beyond this there are many studies of individual road projects and urban LRT projects which are claimed, at least in *ex ante* studies, to be important in the regeneration of economic activity in an area, but for which the evidence is much more mixed in the results from those *ex post* studies which have been carried out.

Project and programme appraisal: These various studies suggest three types of issue which need to be taken into account. *First,* there is the problem of double counting. Conventional transport models are good at estimating traffic on a new link and from this the transport user benefits which are expected to occur. What is more difficult is the identification of what may be termed land use benefits, without the double counting which occurs when for example transport benefits from improved accessibility lead to increased profits and thus land rents at businesses served by the transport infrastructure improvement. If there are any wider benefits, these have to be associated with impacts on productivity, competition or employment. Such effects are likely to be longer term and, as we have seen above, much more difficult to estimate. They need a much wider level of analysis than typically available from a transport model.

Secondly, we face the problems of the two-way road effect and network effects. These require careful definition of the geographical study area for any project to ensure that it is as inclusive as possible of all possible wider effects. It also requires careful definition of the economic information needed, since different projects will have different types of impact on different economic sectors.

Thirdly, the definition of best practice requires that [different types of] appraisal are carried out at both project and programme level in order to ensure consistent treatment of all types of investment.

Land-use transport interaction [LUTI] models: The usual way of treating such issues is through LUTI models, which can be both of standard and extended forms. These require careful definition of three sets of elements: actors, markets and decisions.

Actors are those various groups who take the decisions. These include residents, firms, government, developers, transport infrastructure providers and transport service providers. Each of these has a distinct role in the decision framework and the distribution of benefits between these various groups is likely to be a key issue of interest to government. These actors interact through various markets of interest, property, labour, product and transport. The various actors take decisions in each market relating to the following typical set of questions: whether to, where to, how to and when to do something. Each of these is relevant to the transport outcome.

LUTI models depend for their usefulness on how appropriately the study area is defined and what complexity has been allowed for in the market interactions. They are usually limited by the assumptions made about market behaviour and market interaction, essentially the extent to which perfectly competitive and equilibrium behaviour is assumed. More sophisticated models allow for a greater degree of recursive behaviour and attempt to generate general equilibrium whilst allowing for imperfect competition within individual markets.

Alternatives: We consider two alternative approaches to the modelling of the wider impacts of transport infrastructure investment, computable general equilibrium models and transport linked macroeconomic models. Computable general equilibrium models define all relevant markets and linkages, allowing for product differentiation and imperfect competition in these markets and then compute an equilibrium solution numerically. Their advantage is that they can provide a comprehensive picture of the entire economy which allows for any interactions and feedbacks between all sectors and also allows for sensitivity analysis and policy experiments. The disadvantages of such models is that they can be excessively complex with enormous data requirements. They also depend on simplifying assumptions and find it difficult to deal with induced changes in behaviour since they depend on imported behavioural parameters [elasticities] from outside the model.

Transport linked macroeconomic models use a detailed transport model which is linked as an input to a standard macroeconometric model, for example transport changes are treated as any change in a standard input variable, e.g. productivity. These modes have the advantage that they use existing model shells and emphasise the importance of macroeconomic growth, productivity and employment impacts. However, their disadvantage is that the key linkages are often difficult to specify and they make assumptions about these linkages which involve under-researched theoretical elements.

8.8 Conclusions

In this contribution we have set out some of the approaches which have been used to assess the wider and longer-term impacts of transport infrastructure investment. It has been shown that this is a complex process which demands more than the estimation of a simple aggregate relationship between the infrastructure investment and the level of economic activity. The key elements required are a sufficient representation of both the sectoral and spatial structures of the relevant economy to capture all of the changes which the investment may induce. This is to argue that there is no universal formula which can be applied. The key elements are an appreciation of the competitive structure of the markets affected, the significance of transport costs in both the provision of inputs and the delivery of outputs for that sector [and hence its size of market area] and the scope for substitution of transport for other inputs.

The complexity of this structure is such that most appraisal exercises, for either projects or programmes, will need to use models which simplify these relationships. However, it is important to understand the elements outlined above in order to be able to interpret the findings of any such model. Too frequently projects are justified on no other basis than that it is expected they will have wider economic benefits when a straight transport user benefit appraisal cannot lead to justification for the investment. In some cases, a more thorough analysis of these wider effects will demonstrate that such benefits are at best illusory, at worst actually negative for the area concerned.

References

Aschauer D.A. (1989): Is Public Expenditure Productive? *Journal of Monetary Economics* 23, 177-200

Banister D., Capello R. and Nijkamp P. (eds.) (1995): *European Transport and Communications Networks: Policy, Evolution and Change*, Wiley, London

Batey P.W.J., Madden M. and Scholefield G. (1993): Socio-Economic Impact Assessment of Large-scale Projects using Input-Output Analysis: A Case Study of an Airport, *Regional Studies* 27, 179-191

Batten D. and Karlsson C. (eds.) (1996): *Infrastructure and the Complexity of Economic Development*, Springer, Berlin, Heidelberg, New York

Botham R. (1983): The Road Programme and Regional Development: The Problem of the Counterfactual. In: Button K.J. and Gillingwater D. (eds.) *Transport, Location and Spatial Policy*, Gower, Aldershot, pp. 23-56

Brülhart M. (1998): Economic Geography, Industry Location and Trade: The Evidence, *The World Economy* 21, 775-802

Dodgson J. (1974): Motorway Investment, Industrial Transport Costs and Sub-Regional Growth, *Regional Studies* 8, 75-80

Duffy-Deno K.T. and Eberts R.W. (1991): Public Infrastructure and Regional Economic Development: A Simultaneous Equations Approach, *Journal of Urban Economics* 30, 329-343

European Commission (1997): The Likely Macroeconomic and Employment Impacts of Investments in Trans-European Transport Networks, Commission Staff Working Paper, SEC(97)10, January

Fritsch B. and Prud'homme R. (1997): Measuring the Contribution of Road Infrastructure to Economic Development in France. In: Quinet E. and Vickerman R.W. (eds.) *The Econometrics of Major Transport Infrastructures*, Macmillan, London, pp. 45-67

Gramlich E. (1994): Infrastructure Investment: A Review Essay, *Journal of Economic Literature* 32, 1176-1196

Holtz-Eakin D. (1993): State Specific Estimates of State and Local Government Capital, *Regional Science and Urban Economics* 23, 185-210

Holtz-Eakin D. and Lovely M.E. (1996): Scale Economies, Returns to Variety and the Productivity of Public Infrastructure, *Regional Science and Urban Economics* 26, 105-123

Holtz-Eakin D. and Schwartz A. (1995): Infrastructure in a Structural Model of Economic Growth, *Regional Science and Urban Economics* 25, 131-151

Hotelling H. (1929): Stability in Competition, *Economic Journal* 39, 41-57

Hulten C. and Schwab R. (1991): Public Capital Formation and the Growth of Regional Manufacturing Industries, *National Tax Journal* 44, 121-134

Jara-Diaz S.R. (1986): On the Relations between Users' Benefits and the Economic Effects of Transportation Activities, *Journal of Regional Science* 26, 379-391

Jensen-Butler C. and Madsen B. (1996): Modelling the Regional Economic Effects of the Danish Great Belt Link, *Papers in Regional Science* 75, 1-21

Krugman P. (1998a): Space: The Final Frontier, *Journal of Economic Perspectives* 12, 161-174

Krugman P. (1998b): What's New about the New Economic Geography? *Oxford Review of Economic Policy* 14(2), 7-17

Lau S-H.P. and Sin C-Y. (1997): Public Infrastructure and Economic Growth: Time Series Properties and Evidence, *Economic Record* 73, 125-135

Lynde C. and Richmond J. (1993): Public Capital and Long Run Costs in UK Manufacturing, *Economic Journal* 103, 880-993

Munnell A.H. (1990): How Does Public Infrastructure Affect Regional Economic Performance. In: Munnell A.H. (ed.) *Is There a Shortfall in Public Capital Investment*, Conference Series 34, Federal Reserve Bank of Boston, pp. 69-103

Munnell A.H. (1992): Infrastructure Investment and Economic Growth, *Journal of Economic Perspectives* 6, 189-198

Rietveld P. (1989): Infrastructures and Regional Development, *The Annals of Regional Science* 23, 255-274

Rietveld P. and Bruinsma F. (1998): *Is Transport Infrastructure Effective?* Springer, Berlin, Heidelberg, New York

SACTRA (Standing Advisory Committee on Trunk Road Assessment) (1999): *Transport and the Economy*, Report to Department of the Environment, Transport and the Regions, The Stationery Office, London [Available on-line at www.roads.detr.gov.uk/roadnetwork/sactra/report99/index.htm]

Simmonds D. and Williams I. (1998): Review of Land Use/Transport Interaction Models, Report to SACTRA, David Simmonds Consultancy in Collaboration with Marcial Echenique and Partners

Venables A. and Gasiorek M. (1998): The Welfare Implications of Transport Improvements in the Presence of Market Failure, Report to SACTRA

Vickerman R.W. (1987): The Channel Tunnel: Consequences for Regional Growth and Development, *Regional Studies* 21, 187-197

Vickerman R.W. (1989): Measuring Changes in Regional Competitiveness: The Effects of International Infrastructure Investments, *The Annals of Regional Science* 23, 275-286

Vickerman R.W. (ed.) (1991): *Infrastructure and Regional Development*, Pion, London [=European Research in Regional Science, Vol.1]

Vickerman R.W. (1995a): The Regional Impacts of Trans-European Networks, *The Annals of Regional Science* 29, 237-254

Vickerman R.W. (1995b): Location, Accessibility and Regional Development: The Appraisal of Trans-European Networks, *Transport Policy* 2, 225-234

Vickerman R.W. (1997): High Speed Rail in Europe – Experience and Issues for Future Development, *The Annals of Regional Science* 31, 21-38

Vickerman R.W. (1998): Transport, Communications and European Integration. In: Pinder D. (ed.) *The New Europe: Society and Environment*, Wiley, London, pp. 223-238

Vickerman R.W., Spiekermann K and Wegener M. (1999): Accessibility and Regional Development in Europe, *Regional Studies* 33, 1-15

9 Socio-economic Dynamics and Spatial Mobility: A Scenario Application to Environmental Strategies in Transport

Peter Nijkamp, Sytze A. Rienstra, Richard T.M. Smokers and Jaap M. Vleugel
Free University Amsterdam, Department of Spatial Economics

9.1 Introduction

The 21st century will see a variety of new social and technological trends which will influence the way in which transport is supplied and utilized. At present a wide range of social phenomena, including rising incomes, increased leisure time, new communication technologies, an ageing population and a declining role of the traditional family are changing the nature of the demands we place on transport. In response to new techniques of production, shipping and the growth of markets economic activities are also changing. Institutional reforms such as privatisation and deregulation have changed transport in ways that are not yet well understood. At the same time the increasing use of petroleum resources for travel and transport has raised concerns about the eventual depletion of fossil fuels as well as the contribution to global warming and decreases in urban air quality. Europe and the United States are the world's major energy consumers and transport users. However, the long-term sustainability of current transport systems is increasingly being questioned as levels of motor vehicle fatalities and injuries as well as congestion continue to rise. These trends raise questions on whether our current transport systems will be sustainable beyond the next half century (see also Black and Nijkamp, 2001).

The awareness has grown that in the past decades the industrial, economic and social organization of our societies has exhibited a new form of architecture, in which interaction, communication and association play a prominent role. This new configuration, often coined 'the network economy', has a multi-modal and open structure, in which nodal centres act as the core of the network infrastructure concerned. This applies to both physical and virtual networks.

This new phase in the economic history of the industrialized world provokes three intriguing research challenges:

- Which are the socio-economic, institutional and technological drivers of this modern development?
- Which are the implications of the world-wide transition to a network economy for human behaviour, for the spatial and mobility patterns of human activity and for the industrial organization of our world?
- Which are proper policy responses to this new megatrend in terms of adjusted strategies or new institutions [including private-public arrangements]?

Clearly, the above research questions need a cross-national systematic framework of a multidisciplinary nature. Disciplines to be involved are inter alia economics, management science, sociology, political science, psychology and law, geography and regional science. The study of the network economy is concerned with many dimensions, and hence a dedicated set of focal points in research is necessary.

The present is different from the past, and there is no doubt that the future will be different from the present. But in what respect and to which extent will it be different, and what will be the socio-economic consequences of foreseeable and non-foreseeable changes in the way we are living now? From a transportation point of view our main interest lies in finding answers to questions like: Will future land-use patterns be different from the past? How will mobility and modal split look like in say 30 or 40 years from now? Will current trends continue? What about policy making? And will we find an answer for the seemingly permanent grid-locks in our main infrastructure networks?

In any case, it is clear that the future of transport is fraught with many uncertainties. The impact of transport on society is both positive [due to its contribution to economic development and its role in fulfilling social needs], and negative [e.g., environmental impact of transport]. In economic terms, this negative impact is considerable; recent estimates indicate it they may be as big as three percent of GNP (Verhoef, 1994).

Transport contributes considerably to the emissions of CO_2, which cause, for example, the greenhouse effect. International policy documents indicate that these emissions should be reduced by 50 percent in the year 2025, which may make a reduction in the Western world necessary of about 80 percent [in order to give developing countries the opportunity to increase their emissions to some extent] (WMO, 1988). The reduction of such emissions may be hard to achieve, especially in transport. In the period 1971-1990 in OECD-Europe, for example, the yearly CO_2 emissions caused by the transport sector have increased by 63 percent. The industrial sector reduced its emissions by 27 percent, while the remaining sectors achieved a reduction by 8 percent (calculations based on OECD, 1993). These large emission increases in the transport sector are the result of a continuing rise in the mobility level and a modal shift towards car and airplane, which are the most polluting modes. Efficiency improvements in the same period do apparently not compensate for these trends. It may be concluded, therefore, that far reaching changes are necessary in order to achieve sustainability objectives such as a drastic reduction of CO_2 emission.

This leads to many questions, such as: Which options are available to achieve these changes? What is the role of policy making in implementing these options? What obstacles should be removed before these goals can be reached? How long will it take to reach a new equilibrium in terms of environmental quality?

Scenarios may be an extremely useful tool to analyse possible futures and the impacts of several policy packages. This chapter aims to analyse to what extent scenarios may be useful by first discussing the scenario methodology and next by elaborating some energy scenarios for the future of passenger transport in Western Europe (2030).

The chapter is set up as follows: Section 9.2 introduces the scenario technique, focussing in particular on transportation. Then we will construct energy scenarios for transport systems. In Section 9.3 the energy characteristics of present and likely future transport technologies are assessed. Next, in Section 9.4 we will briefly describe four contrasting base scenarios regarding land use, the institutional framework, economic factors and the individual disposition to change the modal split [including a choice for new or improved modes] and to reduce mobility growth. Section 9.5 shows to what extent future electricity generation and fuel supply systems – as defined by the scenarios – will differ from present systems and what this means in terms of CO_2 emissions. Section 9,6 analyses the differences between present and likely mobility rates and modal shares per scenario. In Section 9.7 these data are combined to calculate overall energy consumption and the resulting CO_2 emissions. Finally, in Section 9.8 we will offer an answer to the main question of this chapter.

9.2 Decision Making and Scenarios

The function of scenarios: At the base of human actions lie [explicitly or implicitly] various choices between two or more [future] options. Decision-making is part of everyday life. Information is the basis and starting-point for these decisions (Samuelson et al., 1977) because it structures decision-making processes. As long as information is valid, it may be assumed that people are able to make proper decisions, and as a result the chance to get proper outcomes [future state, care, etc.] is higher than in the case of 'wrong' information.

In reality, the existing information does often not fulfil the needs of decision-makers in terms of either quantity or quality. This leaves decision-makers, whether they are private households, business firms or government agencies, with quite some degree of uncertainty regarding the likelihood of future developments. As a result, uncertainty largely influences decision-making processes. Uncertainty is always present in open systems, like the transport system. Therefore, it is difficult to predict which developments in these systems are likely in the future and why this is the case. This problem becomes even larger when the time span to be covered is very long [e.g., in this contribution the time span is 35 years].

To reduce these uncertainties, decision-makers need decision support tools. Scenarios are such a tool, which are one of the strategic techniques in future research. Future research is the study field which tries to make statements about the future based on scientific reasoning. The impact and goals of future research may be described as follows (see e.g., Steenbergen, 1983):

- the **signalling** function: scenarios provide better insight into uncertain situations;
- the **communication** and **learning** function: scenarios stimulate thinking about alternative futures;
- the **legitimation** function: scenarios mobilise people and start processes of change;
- the **exploring** and **explaining** function: scenarios show how solutions for specific problems may become reality, given certain policy priorities; they also present possible solution strategies;
- the **demonstration** function: scenarios show the consequences of specific decisions.

The use of scenarios in the transport field: There are various definitions of scenarios. The following definition, which covers all aspects of scenario development may be useful (Steenbergen, 1983, p. 11):

> *'a scenario describes the present situation in [segments of] society, together with likely and desirable future states of this society, and series of events [or transition paths], which may connect the present situation and future states of society'.*

Scenarios enable us to reduce complexity and facilitate discussions about future events by arranging and classifying information and preventing information overload. Ideally, scenario research shows the underlying conditions of various likely futures. Scenarios may help us to gain insight into the consequences of various strategies and enable us to compare them. They may show how small the margins for policy-making are, which is important in situations dominated by external [autonomous] factors (Bijl, 1991). Scenarios may also show the different points of view regarding future choices.

Scenarios cannot be used in all situations. This means that we have to justify their use, in terms of fulfilment of various criteria. In general, scenarios may be used if (Becker and Dewulf, 1990):

- uncertainty about the future is high and historic patterns are shaky; it is very likely that the future will be influenced by events which did not exist before [discontinuities];

- it should be possible for governments to steer the subject;
- there are different opinions about the strategy to be followed [no consensus about means and ends];
- the field should be relatively open [no 'hidden agendas' or blueprints].

When applying these criteria to transport we may observe the following:

- Transport is part of everyday life since humanity exists. Problems in this field, like congestion and air pollution, were already known since the Roman Empire. Transportation history shows the continuous emergence and adoption of new, successive, modes of transport. Each new mode brings new transport potentials to the market, thereby changing the travel opportunities. It is uncertain whether this process will and can also continue in the future. If not, it is very unlikely that the force fields in transport will ultimately force dominant actors to change their use of transport and space dramatically [in a yet unknown direction]. This is also a major cause for concern about the future, which also adds to the existing uncertainty in this field.
- The socio-economic aspects of transportation explain why transportation policy is a well-established part of government policy.
- There are many actors in this field, with diverging opinions about transportation and its future; there is an inherent tension between pursuing allocative efficiency and distributional considerations [equity]. In more general terms, there is the controversy between applying economic principles or administrative ones.
- In modern democracies, the use of blueprints or hidden [political] agendas is prevented by a system of 'checks and balances' in the decision-making process. This does not mean that the future is completely open. Nor is it possible to make drastic changes, as the nature of the decision-making process leads to incrementalism.

A closer look at the historical development of government involvement with modern passenger transportation shows that governments did not well foresee the far reaching consequences [congestion, environment] of the steep growth in car transport. When they did so, they were simply too late, as the car had already established itself strongly in society (Button and Gillingwater, 1986). This is a clear example of the impact of uncertainty in policy-making. It seems as though governments favour running a traditional [certain] course, even if this happens to be a dead-end one.

An important question in this respect is whether the use of scenarios has stopped governments from making the wrong decisions. This is not unlikely, but the problem is that most people are unable to visualise future problems unless they are faced with them in reality. Since this is impossible beforehand, we need an alternative strategic framework, which tells us that certain outcomes of some 'roads to the future' are more or less inevitable. It should be acknowledged that scenarios do not predict that these outcomes should happen and neither how large the effects will be.

Some people have developed theories or 'laws' in this field, such as the law of constant travel time (Hupkes, 1982). Individual time spent on travelling is fairly constant over time because of a stepwise increase of travel speed due to the development of faster vehicles (Grübler and Nakiczenovicz, 1991). In this context, it does not make sense to blame governments for negative impacts of the development of mobility and modal split, since primarily individual motivational factors determine these developments and governments are increasingly restricted in policy-making, even though it is often claimed that governments should use more stringent policy instruments.

We may conclude, that the transport sector shows the above mentioned uncertainties to a large extent. As a consequence, transport is a subject for which scenarios experiments may be a useful tool. In the remainder of this chapter we will carry out such a scenario experiment for the energy consumption and resulting CO_2 emissions in transport.

9.3 Assessment of the Base Scenarios

As a starting point for the scenarios in this contribution we will use the four scenarios constructed by Nijkamp et al. (1998). These scenarios sketch different development paths of the spatial, institutional, economic and socio-psychological determinants of passenger mobility. Two of these scenarios – the market and regulatory scenario – reflect extreme pictures of the future, the other ones — the expected and desired scenarios – are based on empirical expert opinions and mirror more intermediate pictures of the future. In the present contribution the above mentioned – qualitative – scenarios have been used as a starting point for the construction of energy scenarios.

Seventeen European countries are considered: Austria, Belgium, Denmark, Finland, France, Germany, Greece, Ireland, Italy, Luxembourg, the Netherlands, Norway, Portugal, Spain, Sweden, Switzerland and the United Kingdom. For each of these countries the overall transport system [use and modal split] and the energy supply system, the energy consumption and CO_2 emissions are assessed. In the final results the figures of individual countries are added up to obtain European averages.

It should be emphasised that the various estimates are tentative. Input data describing overall future passenger mobility and the future composition of the energy system in the four different scenarios are dependent on the assumptions made, although available studies and existing forecasts are used to the maximum extent possible. An advantage of this approach is that several energy consumption scenarios can be used and tested. The same may hold for different patterns of mobility levels. As a consequence, the results are mainly intended to give an indicative picture of the possible impacts on sustainability as a result of the developments described in the various scenarios. These results aim also at assessing the possible impacts of several energy policies. First, we will now briefly describe the main features of and assumptions made in the four scenarios.

The market-oriented reference scenario: The market-oriented reference scenario assumes that the government largely withdraws from the economy and the society. When the government intervenes, this is done in a market-based way. Trends like individualisation and deregulation are even reinforced, while the spatial organisation becomes very diffuse. The resulting transport system is almost entirely based on individual modes. At the same time new fuels [electricity, liquid hydrogen] are introduced in almost all cars.

The regulatory reference scenario: This scenario is largely the opposite of the market-oriented scenario. Here current trends are mostly reversed, while the spatial organisation becomes concentrated and compact. The government stimulates collective modes, while car use and air transport are largely discouraged. The transport system is almost entirely based on collective modes.

The expected expert-based scenario: In the expected scenario, trends and underlying factors will not significantly change in the future. For example, no large scale behavioural changes are expected for individual citizens in order to reduce the externalities caused by transport. Also trends like individualisation and deregulation will largely continue. The same holds for the government, whose policy is not expected to be sufficiently sustainable. Only spatial policy may be rather successfully implemented. New transport modes and technologies, like new fuels [e.g. hydrogen] or collective modes, will not be introduced at a large scale. In conclusion therefore, the transport system and society may more or less resemble today's pattern. Also the mobility growth will largely continue.

The desired expert-based scenario: In the desired scenario, a clear choice is made for a collective transport system. Therefore, far reaching changes are necessary in the behaviour of individuals, as well as in the institutional and economic environment. For example, large scale regulation is necessary, while the individualisation trend should largely stop. For example, the government should largely invest in public transport infrastructure and take drastic measures to reduce car use in order to ensure that mobility growth will largely slow down. It may be clear that it will be very hard for the government and for society to let this scenario become reality. One of the most pronounced practical problems that will emerge in society is free riding.

Next, these scenarios will be used as a starting point for the energy scenarios. The way the analysis is carried out in the rest of this chapter is presented in Fig. 9.1.

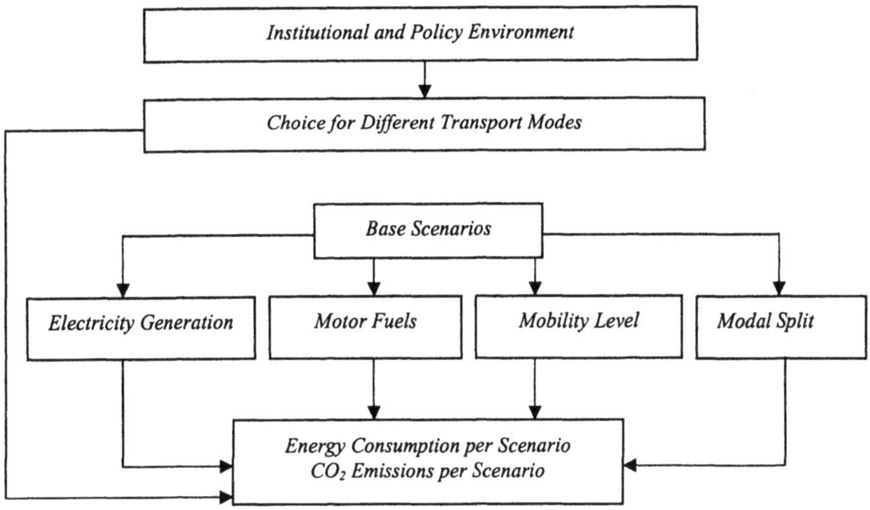

Fig. 9.1 Structure of the analysis

9.4 Energy Consumption of Present and Future Transport Modes

The first step in calculating the CO_2 emissions of transport modes is to estimate the energy consumption in passenger transport, which is of course related to spatial structures, mobility patterns and modal choices. In the scenarios, a distinction is made between urban and long-distance passenger transport. The energy consumption of transport modes is expressed in terms of the energy content of final energy carriers [electricity or motor fuels like gasoline, diesel, biofuel, kerosine or hydrogen] consumed per passenger kilometer.

Table 9.1 Final energy consumption of various transport modes (1989, MJ/pass. km)

Car		Bus		Train		Plane
Urban	Long Distance	Urban	Long Distance	Urban	Long Distance	Long Distance
1.70	1.17	0.75	0.60	0.35	0.35	2.00

Notes: MJ denotes megajoule, pass. km passenger kilometers

Sources: Smokers and van Hilten (1992); IEA/OECD (1993); confidential ECN-data

Table 9.2 Expected final energy consumption of transport modes (2030, MJ/pass. km)

	Urban Transport	Long-Distance Transport
Car		
Internal Combustion Engine Vehicle		
[Conventional]	0.90	0.70
[Alternative Fuel]	0.90	0.70
Hybrid-Electric Car	0.50	0.55
Battery-Electric Car	0.40	0.45
Bus		
Internal Combustion Engine Vehicle		
[Conventional]	0.50	0.55
[Alternative Fuel]	0.50	0.45
Hybrid-Electric Car	0.40	0.35
Fuel Cell	0.40	0.30
Battery-Electric Car	0.20	
Train	–	0.30
High Speed Train	–	0.35
Plane		
Conventional	–	1.40
Liquid Hydrogen	–	1.20
Tram	0.25	–
Metro	0.25	–
People Mover	0.30	–

Notes: MJ denotes megajoule, pass. km passenger kilometers

As the estimates of a specific technology often vary from source to source, the data presented in this section are given in general averages, weighted on the basis of expert judgement and expert estimations. In various cases the specific numbers cannot be related to a single literature source, but to a blend of various data sources and expert knowledge at ECN (Energy Centre, The Netherlands) in Petten, and are therefore presented without an unambiguous reference.

Due to the aggregated nature of the available European transport data, only four transport modes are discerned in the transport system of 1989: the car, bus, train and plane. Table 9.1 provides the final energy consumption factors of these modes for urban and long-distance passenger transport in 1989. For the scenarios in our study a larger set of transport technologies has to be taken into account. Table 9.2 presents the estimated final energy consumption of these transport modes in the year 2030, in which five different passenger cars and buses are considered in the calculations (see also Nijkamp et al., 1998):

- a very energy efficient version of the conventional car [internal combustion engine vehicle (ICEV)];
- an ICEV with a biomass-derived alternative fuel [e.g. methanol];

- a hybrid-electric car [HEV], with an electric drive system and a gasoline-fuelled combustion engine operating as generator;
- a fuel cell powered car [with hydrogen as energy carrier];
- a battery-electric car [BEV] [only for urban transport].

In the scenario pictures for urban public transport also trams, metros and people movers may be used, which are all electrically driven. Collective long-distance transport too may be performed by electrically propelled conventional trains, high-speed trains [HST] and trains with [electro-]magnetic levitation and propulsion [Maglev]. Planes may be conventionally [kerosine] fuelled, or liquid hydrogen [LH_2] fuelled. Although the technological development of the various future transport modes will differ to some extent in the scenarios, it has been decided to use the same energy consumption figures for every mode in each scenario.

9.5 Electricity and Motor Fuels

After the description of the various modes in terms of energy consumption it is necessary to investigate which fuels [primary and final sources] are used in the above discussed scenarios, while also the emissions of these energy sources in the year 2030 have to be estimated. First, we will discuss the electricity generation as well as the assumptions made in the calculations for the scenarios. Next, the figures on future motor fuels will be dealt with.

Electricity generation: In the scenarios different assumptions will be made for the generation of electricity. These assumptions have large impacts on the final CO_2 emissions. In order to calculate the primary energy consumption of electrically driven modes, their final energy consumption factor must be divided by the efficiency with which it is generated. In a similar way, the resulting CO_2 emission per passenger kilometer is calculated by multiplying the final energy consumption with a CO_2 emission factor. The efficiency of the electricity generation system is calculated by dividing the yearly electricity output by the yearly input of primary energy carriers. Similarly, the CO_2 emission factor is defined as the total yearly CO_2 emission from the electricity generation divided by the total yearly electricity production. The conversion efficiencies and CO_2 emission factors of electricity generation from coal, gas, oil, uranium and renewables are presented in Table 9.3.

For each of the 17 European countries mentioned above, the composition of the electricity generation system is projected for the scenarios concerned. The input for the expected scenario is based on the average extrapolated European energy data for the year 2020, as estimated by Oostvoorn and Pellekaan (1995).

The composition of the electricity generation in the other three scenarios is derived from that of the expected scenario in an informal way, taking into account the distinctive characteristics of the institutional, economic and socio-psychological determinants in the scenarios. We have attempted to create four distinctively different energy scenarios in order to broaden the analysis and to analyse the sensitivity of the outcomes. The resulting shares of various energy sources in the European electricity generation mix are presented in Table 9.4.

Motor fuels: Table 9.5 shows some characteristics of the motor fuels that are taken into account in the estimations for 1989. The first two columns present the efficiency of fuel refining as units of final energy input divided by units of primary or fossil energy input, which are equal for all fossil fuels. The third and fourth column give the CO_2 emission per unit of primary and final energy consumed, the last column the CO_2 emission per unit of fuel energy content. In the CO_2 emission per unit of final energy consumed the indirect emissions resulting from oil refining have been incorporated too. Similar characteristics of the motor fuels related to the year 2030 are presented in Table 9.6.

Table 9.3 Characteristics of processes for electricity generation

Energy Source	**Coal**	**Gas**	**Oil**	**Nuclear**	**Renewable**
g CO_2 per MJ Prim	94	56	74	0	0
1989					
Efficiency (%)	36	37	36	33	100
g CO_2 per MJ Fin	260.39	152.59	208.45	0.00	0.00
2030					
Efficiency (%)	50	58	40	33	100
g CO_2 per MJ Fin	188.00	96.55	185.00	0.00	0.00

Notes: Prim denotes primary energy consumption, Fin final energy consumption

Table 9.4 Share of various energy sources [percent energy input]

			Source		
Scenario	**Coal**	**Gas**	**Oil**	**Nuclear**	**Renewable**
1989	35	6	9	38	12
Market-Oriented	22	16	7	27	28
Regulatory	8	10	3	48	31
Expected	30	7	13	38	12
Desired	24	13	5	21	37

Table 9.5 Characteristics of various motor fuels (1989)

	MJ Fin/ MJ Prim (%)	MJ Fin/ MJ Fossil (%)	CO_2 Prim (g/MJ)	CO_2 Fuel (g/MJ)	CO_2 Fin (g/MJ)
Gasoline	90	90	74.50	74.50	82.78
Diesel	93	93	73.00	73.00	78.49
Kerosine	90	90	73.00	73.00	81.11

Notes: Prim denotes primary energy consumption, Fin final energy consumption

Table 9.6 Characteristics of various motor fuels (2030)

	MJ Fin./ MJ Prim (%)	CO_2 Prim (g/MJ)	CO_2 Fuel (g/MJ)	CO_2 Fin (g/MJ)
Gasoline	90	74.50	74.50	82.78
Diesel	93	73.00	73.00	78.49
Alternative Fuel*	55	0.00	70.00	0.00
Fuel Cell	100	0.00	0.00	0.00
Kerosine	90	70.00	70.00	77.78
Liquid Hydrogen	100	0.00	0.00	0.00

Note: * Bio-Ethanol

Besides gasoline, diesel and kerosine, two alternative motor fuels have been considered in our analysis. These are a biomass-derived fuel [e.g. bio-ethanol or biodiesel] and hydrogen. In road transport hydrogen is assumed to be used in fuel cells. Hydrogen can also be burnt in the gas turbines of planes, which carry the fuel in liquid, cryogenic form. The efficiency of the production of alternative fuel from primary biomass is relatively low. It is assumed that all energy requirements in the energy chain of biomass derived fuel are covered by energy from biomass. The efficiency of fossil to final energy conversion is, of course, infinite for biomass-derived alternative fuel and hydrogen, but has been set to a sufficiently high finite number in the model calculations.

9.6 Developments in Passenger Mobility

For the estimation of the CO_2 emissions also the mobility level in the distinct scenarios has to be determined. This will be dealt with in the next subsections. First, the mobility level in 1989 is investigated. Next, the assumed mobility levels in the distinct scenarios are discussed.

The mobility level in 1989: The passenger mobility and modal split in European countries in 1989 has been calculated by using ECMT (1993) and Eurostat (several years) data. For Greece, Ireland and Luxembourg complementary data have been found in Samaris and Zierock (1992). Available statistics do not make a distinction between passengers in urban and long-distance transport, however. Since such a distinction is necessary in the scenarios, the division of total mobility into urban and long-range transport has been estimated by taking the trips below 10 kilometers as urban and above 10 kilometers as long-distance transport and by estimating next their average length. Also a European average has been applied to all countries. Although this assumption may be rather crude, it is necessary to include all passenger kilometers in the analysis. Moreover, the results presented in Table 9.7 look rather plausible, and in any case this assumption is not expected to have a large impact on the final results. The total European passenger mobility in 1989 amounted to 3948×10^9 passenger kilometers of which 27 percent is assumed to be urban transport and 73 percent long distance transport.

Starting at this point, the mobility level in the scenarios can now be estimated. This will be elaborated by extrapolating current trends and by interpreting and discussing the assumptions made in the successive scenarios.

Table 9.7 Disaggregation of passenger transport in Europe by mode and distance (1989)

	Car		Bus		Train		Plane
	Urban	Long Distance	Urban	Long Distance	Urban	Long Distance	Long Distance
European Average	29	71	50	50	10	90	100

Trends in mobility: In most studies about future transport scenarios a simple model is used, in which current trends in specific parameters are translated into the future by means of a linear multiplier. This assumes that past trends may be used as an indicator for future trends. If we look at these developments in Europe, the expectations may be as the ones presented in Table 9.8.

If the past trends are to be regarded as a blueprint for the future, mobility will apparently explode. However, one may doubt whether this will actually happen because in these trends it is assumed that:

- there is no restriction in the capacity of the infrastructure; in reality however, unlimited infrastructure expansion is not realistic;
- the time budget of individuals is unlimited; in practice, however, this is not realistic [of course, faster modes may be introduced] (Hupkes, 1982);
- car ownership should grow at the same rate as in the last 25 years; however, this neglects the existence of saturation;
- population growth should continue at the same pace.

Table 9.8 Mobility trends in Western Europe [excluding Luxembourg, Greece and Ireland]

	Average Growth Rate 1965-1989 (%)	Index 2030 (1989 = 100)
Total	3.7	428
Car	4.3	539
Train	1.1	155
Bus	1.3	168
Air	8.1	2254

Note: Only passenger kilometers by private car, rail and bus are taken into account
Sources: Calculations based on ECMT (1993) and Eurostat (several years)

There are no applicable population forecasts available for 2030. The major reason for this is the way in which population data are gathered in the various European countries. For the EU, Eurostat (1993) includes, for example, a forecast for only seven countries for the year 2020. These indicate that the population will stabilise at the beginning of the next century, which is mainly due to the low birth rates. For the above-mentioned reasons it may be expected that current mobility growth will continue at lower growth rates. Therefore, in order to find more plausible growth rates, we will discuss some Dutch forecasts, which may be indicative for the average European growth rates.

According to the official Dutch transport policy memorandum (Tweede Kamer, 1990), car mobility for the year 2010 will rise to some 70 percent above the 1986 level [without additional restraining policy measures]. This means an average growth of 2.15 percent per year, which is considerably lower than the actual growth in the period 1965-1989 [4 percent per year], which clearly followed the European trend. For these mobility growth predictions no separate distinction can be made per mode. Also a more recent study by the Dutch CPB (1992) comes to about the same conclusion. Based on the foregoing observations and assuming no drastic policy changes –only some policy measures to reduce mobility growth – we may expect some reduction in the growth of mobility for the period up to the year 2030.

Table 9.9 Growth indices in passenger transport (1989 = 100)

Scenario	**Market**		**Regulatory**		**Expected**		**Desired**	
	Urban	**Long Distance**	**Urban**	**Long Distance**	**Urban**	**Long Distance**	**Urban**	**Long Distance**
European Average	125	200	185	130	185	225	150	185

The mobility rates in the scenario: By taking the above considerations into account, we have estimated the mobility level in the distinct scenarios. Table 9.9 presents the growth indices of urban and long-distance passenger transport in the different scenarios distinguished (1989 = 100). The assumptions made in these scenarios will be discussed next.

The mobility level in the market-oriented scenario: In the market-oriented reference scenario mobility growth is mainly concentrated in the high income category, which consists of 20 percent of the population, owns a private car and mainly lives in peripheral regions. Also some mobility growth in the middle income class is likely to occur. When we analyse the assumptions made in Nijkamp et al. (1998), the following picture emerges. In long-distance transport, mobility growth in the 2000s is initially negative. If we assume that in the 1990s the growth in mobility was three percent per year, then the 2000 mobility level may be 40 percent above the 1989 level. After the year 2000, mobility growth is reduced to 30 percent above this level. After 2010 mobility grows mainly without restrictions, though it is more expensive than in the 1990s. Therefore, the growth rates are expected to be lower and to rise in the year 2030 to more than 100 percent above the 1989 level, which means an average growth rate of 1.7 percent per year. The urban picture differs considerably from the long distance one. Although the 2000 level equals that of long-distance traffic, this level will stabilise afterwards because of the decreasing population in urban areas. Therefore, this level is expected to be 25 percent higher than the 1989 level [i.e., a growth rate of 0.5 percent per year]. For the other scenarios a similar analysis has been carried out.

The mobility level in the regulatory scenario: In the second reference scenario, all major government policies aim at redressing mobility growth by regulatory measures. The decline in individual mobility is offset by the increase in collective transport. In long-distance transport, the overall mobility level may be relatively low because of the shift towards urban areas and the lower quality [more time-consuming] transport modes. In the 2000s, the mobility level even declines because of the policy measures implemented by then. Therefore, this level may in 2030 be some 30 percent above the 1989 level, which implies a growth rate of 0.6 percent per year. Because of the shift back to the urban areas and the relatively high quality of the collective modes, the mobility level will grow much more in the urban areas, viz. 85 percent [on average 1.5 percent per year].

The mobility level in the expected scenario: In the expected expert scenario, it is taken for granted that in the period up to the year 2030 mobility growth will continue at a lower pace. It may be assumed that long distance mobility will grow at a rate of two percent per year, which means that the 2030 mobility level lies 125 percent above the 1989 level. Also in the urban areas we will observe a considerable mobility increase, which is, however, lower than in long-distance transport, because of the development of a moderate compact city, which reduces transport needs. Therefore, the growth rate is 1.5 percent per year, which means that the mobility level rises to 85 percent above the 1989 level. In total, the resulting mobility growth is lower than in the official Dutch future predictions, because it is assumed that the reduction in growth rates will continue.

The mobility level in the desired scenario: In the desired expert scenario the mobility growth rates are lower than in the expected case, viz. 1.2 percent per year, which is, however, higher than in the regulatory scenario. At the urban level the compact city concept is introduced in this scenario. After the introduction of anti-car measures, mobility declines quite strongly, but later the level will rise again due to economic and population growth and the occurrence of new services. Therefore, the former growth rate is re-established. In total, this results in a mobility level of 50 percent above the 1989 level. This is lower than in the regulatory scenario, because in the latter scenario the urban population growth is lower than in the desired scenario. In the broader European mobility analysis, identical growth rates are assumed for all countries. By applying these growth rates to the total urban and long-distance mobility of individual countries in 1989, the urban and long-distance mobility in the scenarios can be calculated. The total passenger mobility in 1989 and for the scenarios for 2030 is depicted in Fig. 9.2.

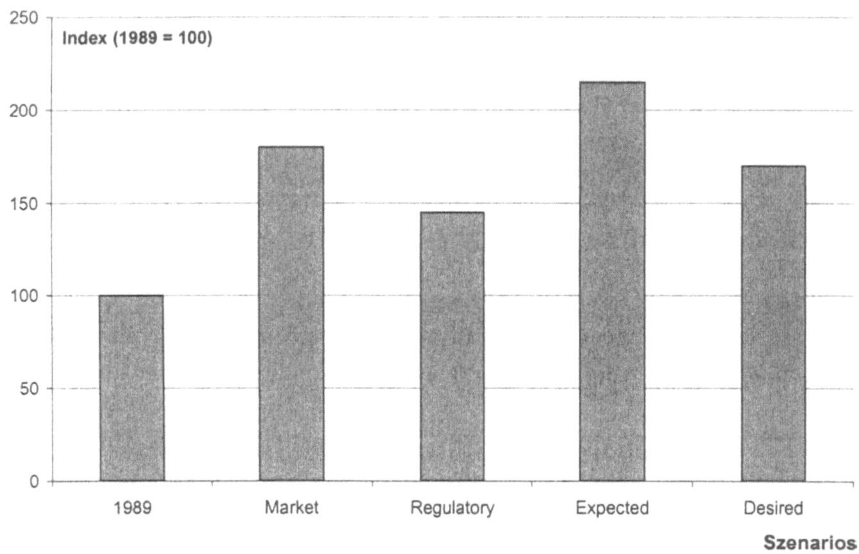

Fig. 9.2 Estimated mobility level in the scenarios (index: 1989 = 100)

The future modal split including new technologies: Given the overall mobility data for urban and long-distance passenger transport in the scenarios, we will now describe the implementation of the various transport technologies. The characteristics of the base scenarios described in Section 9.3 are used as basis for composing a plausible modal split for each scenario in 2030. When a mode is supposed to almost disappear or not to be introduced in a scenario, the modal share is assumed to be '0'. In this way the scenarios have been made more distinct from one another in order to create a clearly focused the scenario analysis. The assumed modal splits in terms of the shares of various technologies in urban and long distance transport are presented in Tables 9.10 and 9.11.

Both reference scenarios present a clear modal split in favour of either the car or collective modes, while the expert scenarios are somewhat less extreme. The modal share of the plane remains low in each scenario, a situation also indicated in the scores for the distinct modes. For the modal split between car types, the assumptions made in Section 5.1 for electricity generation are taken into account. All these data are now used to assess the energy consumption for different scenarios.

Table 9.10 Urban modal split in the scenarios (%)

	Scenario			
Mode	**Market**	**Regulatory**	**Expected**	**Desired**
Car				
Internal Combustion Engine Vehicle				
[Conventional]	0	0	20	0
[Alternative Fuel]	35	0	10	10
Hybrid-Electric Car	0	0	0	0
Battery-Electric Car	40	10	10	10
Total Car	75	10	40	20
Bus				
Internal Combustion Engine Vehicle				
[Conventional]	0	0	15	0
[Alternative Fuel]	10	0	0	0
Hybrid-Electric Car	0	10	10	10
Fuel Cell	5	15	5	10
Battery-Electric Car	10	10	10	10
Tram	0	20	10	20
Metro	0	20	5	20
People Mover	0	10	0	10
Train	0	5	4	0
Total Modes	25	90	60	80
Total	100	100	100	100

Table 9.11 Long distance modal split in the scenarios (%)

	Scenario			
Mode	**Market**	**Regulatory**	**Expected**	**Desired**
Car				
Internal Combustion Engine Vehicle				
[Conventional]	0	0	30	0
[Alternative Fuel]	90	20	10	10
Hybrid-Electric Car	0	0	10	0
Fuel Cell	0	0	0	10
Total car	90	20	50	20
Bus				
Internal Combustion Engine Vehicle				
[Conventional]	0	0	1	0
[Alternative Fuel]	5	0	1	5
Hybrid-Electric Car	0	0	3	0
Fuel Cell	0	0	0	15
Train	0	60	27	40
High Speed Train	3	15	8	15
Trains with [Electro-] Magnetic Levitation and Propulsion	1	5	0	0
Total Modes	9	80	40	75
Plane				
Conventional	0	0	10	0
Liquid Hydrogen	1	0	0	0
Total	100	100	100	100

9.7 Analysis of Results

General results: The total number of passenger kilometers in each of the European countries has been calculated for all transport technologies in 1989 and 2030. By using the above data, the total consumption of final, primary and fossil energy and the CO_2 emissions by European passenger transport in 1989 and in the scenarios for 2030 can now be calculated [see Fig. 9.3 and Fig. 9.4]. The big difference in fossil energy use between the expected scenario and the other scenarios is striking. The same holds for CO_2 emissions [see Fig. 9.4].

Fig. 9.5 presents the relative consumption of various primary energy carriers by European passenger transport in 1989 and in the scenarios for 2030. Actually, the hydrogen [H_2] listed in Fig. 9.5 is not a primary energy carrier. In all scenarios hydrogen is assumed to be produced from renewable energy sources [e.g. photovoltaic] with 100 percent [virtual] efficiency, however. The contribution of hydrogen fuel has been split off here from the total contribution of renewables in order to be able to distinguish it from the use of renewables through electricity consumed by electrically propelled transport systems. The results for the individual scenarios will be described and analysed in the next subsection.

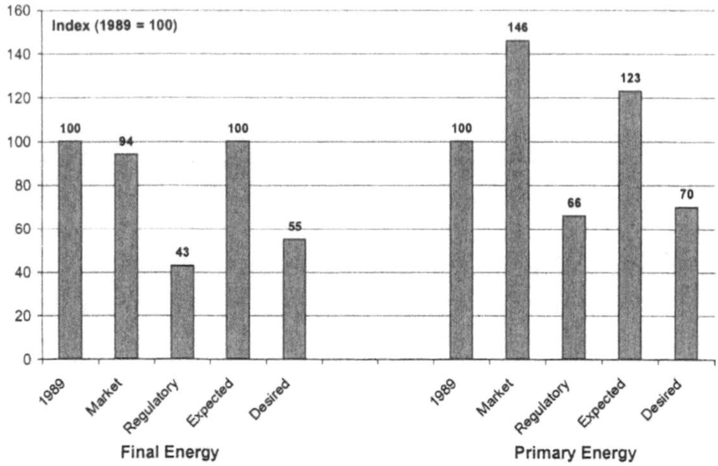

Fig. 9.3 Use of final and primary energy in the scenarios (index: 1989 = 100)

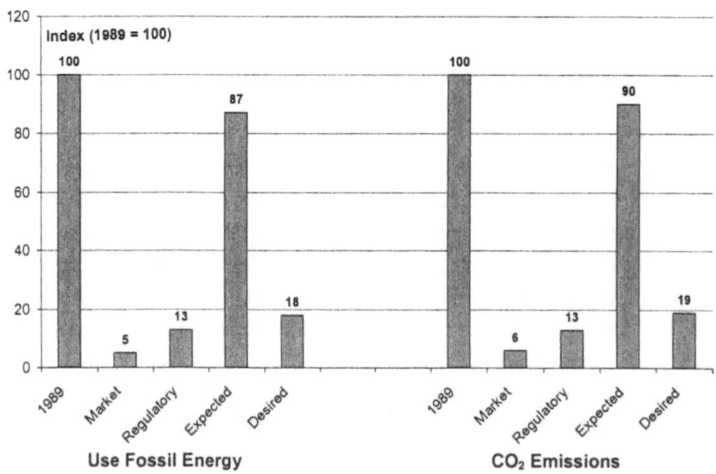

Fig. 9.4 Use of fossil energy and the CO_2 emissions in the scenarios (index: 1989 = 100)

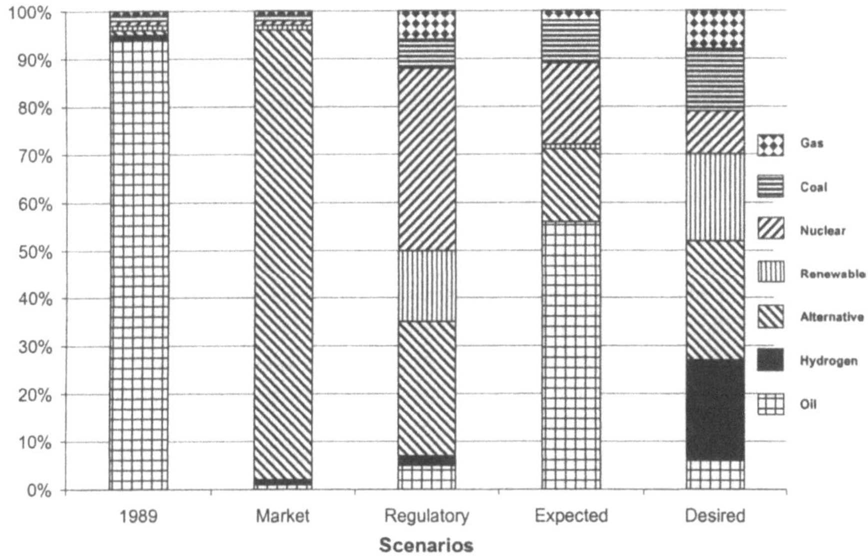

Fig. 9.5 Shares of primary energy sources in the scenarios

Results of the distinct scenarios

The market-oriented scenario: Fig. 9.5 shows that in the year 1989 the passenger transport system almost completely relied on the use of oil as a primary energy carrier. This is, of course, the result of the dominant position of the passenger car. In the market scenario this dominant position of individual road transport has not changed. From Table 9.10 and Table 9.11 it can be seen that the car accounts for 75 percent of all urban passenger transport and 90 percent of all long-distance transport. The use of oil however, has almost completely been substituted for biomass-derived motor fuels. According to Okken et al. (1993a) and Hall (1994) the total potential for biomass in Europe is rather high, so that this scenario stretches the use of biofuels to their ultimate limits. Despite the large mobility increase [82 percent], the final energy consumption is comparable to that in 1989, due to the implementation of ICEVs which are significantly more energy efficient than they were in 1989. The total primary energy consumption in the market scenario is however, larger than in 1989. This is due to the relatively poor conversion efficiency with which motor fuels can be derived from raw biomass [see Table 9.3]. The consumption of fossil fuels and the consequent emission of CO_2 is however, very low in this scenario [95 percent below the level of 1989].

The regulatory scenario: The regulatory scenario combines a relatively slow increase of overall mobility [46 percent] with a dramatic shift in the modal split towards collective transport modes. In total 90 percent of all urban passenger transport and 80 percent of all long-distance transport is assumed to be performed by means of collective transport modes. The calculations show that for this scenario final energy demand can be reduced by more than 50 percent, compared to 1989. The primary energy demand is reduced by only some 30 percent. The shift to collective modes implies a shift to electricity as final energy carrier. In this scenario it has been assumed that a large part of the electricity is generated in nuclear power plants, which have a relatively low conversion efficiency resulting in a rather high primary energy consumption. As however, electricity in the regulatory scenario is for the largest part produced from non-fossil energy sources [nuclear and renewables], the resulting total fossil energy consumption and CO_2 emission is very low [about 88 percent below the level of 1989].

The expected scenario: Mobility in the expected scenario is the highest of all scenarios [115 percent above the level of 1989]. Despite efficiency improvements in transport technologies the final energy demand in this scenario is comparable to that of 1989. Although much less than in 1989, the transport system is still dominated by individual transport by means of passenger cars [about 50 percent of total mobility]. For the most part these are still powered by fossil motor fuels, but there is also some introduction of biofuels and electricity. Collective transport relies on buses, which are mainly diesel powered. Fossil energy consumption and CO_2 emissions in the expected scenario are therefore high compared to the other scenarios, although slightly lower than in 1989. Primary energy consumption is higher than in 1989, but lower than in the market-oriented scenario.

The desired scenario: To some extent the desired scenario is similar to the regulatory scenario. Both are characterised by a significant shift from individual to collective transport modes, although this shift is less drastic in the desired scenario. Also the use of renewable energy sources for electricity production and biofuels for transport is largely comparable. An important difference however, lies in the fact that the use of nuclear energy has been significantly reduced. In the desired scenario, this is partly compensated for by the extensive use of hydrogen, and partly by using electricity from conventional primary sources [coal and gas]. With a figure of 77 percent, the mobility increase is also much larger than in the regulatory scenario. All in all, this still results in a significant decrease in final and primary energy consumption, compared to 1989. The fossil energy consumption and CO_2 emissions are slightly higher than in the market and regulatory scenario, but still 80 percent lower than in 1989.

9.8 Conclusions

Decision-making in transport is fraught with many difficulties. The major obstacle is uncertainty regarding future developments. The combination of the incremental nature of public decision-making and various uncertainties regarding the future, leads many decision-makers to sticking to the same road as they did in the past. Yet, new problems and opportunities demand new tactics and tools. What is needed above all is the awareness that things can be changed, irrespective of how severe today's problems [still] may seem.

Scenarios are an important decision support tool for decision-making in transport. The use of scenarios may help governments to become aware of the impact of specific developments in society in terms of inter alia social costs and the need for public intervention.

From the scenario analysis presented in this chapter, it appears that large-scale changes in the energy composition of future passenger transport systems are possible. On the one hand, these may the result of an increasing use of electricity [in both collective modes and cars]. On the other hand, these shifts may be found in the introduction of non-fossil fuels (see also Deluchi, 1993; Michaelis, 1995). A large variety of new energy sources and technologies may become available in the next century.

However, when the expected scenario is analysed, both collective modes and alternative fuels are not introduced on a large-scale, while at the same time mobility growth continues in the same manner. Emission reductions are mainly the result of efficiency increases of current modes. It is evident that more drastic changes are necessary to arrive at a more sustainable transport system. Current policies do not seem to have an answer on these challenges.

When governments want to achieve a more sustainable transport system, clear policy choices have to be made. It should be acknowledged however, that this may impact on economic development, while also the freedom of choice may be affected [especially in the regulatory scenario]. Also trends like individualisation or the provision of public transport may be affected by a drastic change in policy. This may hold in particular for the spatial organisation of activities and population.

It should thus be recognised, that policies cannot provide an answer to all these problems [at the same time]. Free-rider behaviour, for example, may cause problems, especially when the individualisation trend continues. Many measures only seem to be useful at a European level [for example, a CO_2 tax], because otherwise the competitive position of individual countries may be affected [at least in the short term] and activities may shift to countries with a milder policy climate. Also economies of scale may be achieved when measures are taken at the European level [e.g. in telematics and the development of new engines].

From one scenario analysis, it may thus be concluded, that a more environmentally-friendly transport system is certainly feasible, but this requires large-scale changes in both public policy and individual behaviour.

References

Becker H.A. and Dewulf G. (eds.) 1990: *Terugkijken op Toekomstonderzoek*, RUU-ISOR, Utrecht

Bijl R.V. (1991): *Constructie van Toekomstscenario's*, PhD thesis, RUU, Utrecht

Black W. and Nijkamp P. (eds.) (2001): *Social Change and Sustainable Transport*, Indiana University Press, Indiana

Button K.J. and Gillingwater D. (1986): *Future Transport Policy*, Croom Helm, London

CPB (Central Planning Bureau) (1992): *Verkeer en Vervoer in Drie Scenarios*, Working Document 35, The Hague

DeLuchi M.A. (1993): Greenhouse-Gas Emissions from the Use of New Fuels for Transportation and Electricity, *Transportation Research* 27A (3), 207-216

ECMT (European Conference of Ministers of Transport) (1993): *Annual Trends in Transport 1965-1989*, Paris

Eurostat (1993): *Demographic Statistics 1993*, CEE, Luxembourg

Eurostat (several years): *Basisstatistieken van de Gemeenschap*, Luxembourg, Brussels

Grübler A. and Nakicenovicz N. (1991): Evolution of Transport Systems: Past and Future, IIASA Report RR-91-8, International Institute for Applied Systems Analysis, Laxenburg

Hall D.O. (1994): Biomass Energy Options, IEA Conference on *'Energy Technologies to Reduce CO_2 Emissions in Europe: Prospects, Competition, Synergy'*, Petten, April 11-12, 1994

Hupkes G. (1982): Toekomstonderzoek, Scenario's voor Verkeer en Vervoer, *Planologische Verken-ningen* 4, VUGA Publishers, The Hague

IEA/OECD (1993): *Cars and Climate Change*, IEA/OECD Energy and Environment Series

Leinbach T.R. (2000): Mobility in Development Context, *Journal of Transport Geography* 8, 1-9

Michaelis L. (1995): The Abatement of Air Pollution from Motor Vehicles: The Role of Alternative Fuels, *Journal of Transport Economics and Policy* 29 (1), 71-84

Nijkamp P., Rienstra S.A. and Vleugel J.M. (1998): *Transportation Planning and the Future*, Wiley, New York

OECD (1993): *OECD Environmental Data: Compendium 1993*, Organisation for Economic Cooperation and Development, Paris

Okken P.A., Kram T., Lako P., Ybema J.R. and Gerbers D. (1993): *Kansrijke Energietechnieken: Vier Blauwdrukken voor een Energiehuishouding met Minimale Emissies op Langere Termijn*, ECN Policy Studies, ECN-C-93-036, Petten

Oostvoorn F. van and Pellekaan W. (1995): *Energy Scenarios for a Changing Europe*, ECN, Petten (confidential)

Samaris Z., and Zierock K.H. (1992): *Assessment of the Effect in EC Member States of the Implementation of Policy Measures for CO_2 Reduction in the Transport Sector*, Report for DG XI-B/4

Samuelson K., Borko H. and Amey G.X. (1977): *Information Systems and Networks*, North-Holland, Amsterdam

Smokers R.T.M. and Hilten O. van (1992): *Elektrische Auto's in het Perspectief van de Milieu- en Energieproblematiek*, ECN Policy Studies, ECN-RX--92-073, Petten

Steenbergen B. van (1983): *In de Proeftuin van de Samenleving*, PhD Thesis, RUU, Utrecht

Tweede Kamer (1990): *Tweede Structuurschema Verkeer en Vervoer; Deel d: Regeringsbeslissing*, Meeting Year 1989/1990, 20 922, no. 15-16, SDU, The Hague

Verhoef E.T. (1994): External and Social Costs of Road Transport, *Transportation Research* 28A (4), 273-287

WMO (1988): Implications for Global Security, *Proceedings of the World Conference on the Changing Atmosphere,* Toronto 1988, WMO Report 710, Geneva

PART D: Regional Development and Policy Issues

10 Some Effects of Regional Differentiation on Integration in the European Community*

Gündüz Atalik
Istanbul Technical University, Department of Urban and Regional Planning

10.1 Regional Differentiation and Integration

Disparity, in its specific meaning, can be defined as inequality, difference, or incongruity. Meanwhile, as Myrdal (1972, p. 17) pointed out, '... *experience shows that if, by some chance, discrimination in a particular field of social contact is increased or decreased, the psychological force behind it, that is prejudice, tends to change so as to support actual behaviour. This too fits into the general pattern of circular causation.*' In this sense, there are certain periods when opposing forces balance one another. However, when the whole system moves after a shock by a push or a pull, then the changes in the forces work in the same direction because the variables are interlocked in circular causation.

In fact, regional disparities in the European Union [EU] are obvious, and are attributable equally to the international component and to the regional disparities within the countries. After ten years of the regional development programme there were enormous disparities between the Member States in 1985, both by unemployment rate and by per capita GDP. Although there was considerable overall development for the whole area, no fundamental change could be realised in regional disparities (Knox, 1984; Sant, 1987; CEC, 1984; Eurostat, 1988). This result should give an argument for the theory of comparative advantage when the related factors work for interests in the developed areas.

With respect to the extent of national and regional disparities and the dynamics of regional change such as structural economic change, corporate reorganisation, the decentralisation of economic and social life, the extension of government influence and political integration, Knox (1984) has verified that certain differentials between nations remain substantial. However, at the regional level, contrasts are greater and the pattern of disparities is more sharply defined.

* Reprinted [with slight modifications] from *The Papers of the Regional Science Association* 69, 1990, pp.11-19

In fact, the regional policies given by the Treaty of Rome require the member countries '*to ensure their harmonious development by reducing the differences between the existing regions.*' In this way, EU policies have been formalised as follows: first, to create an awareness of regional problems by organising related efforts such as conferences; second, to coordinate regional measures for the various member countries to avoid wasteful rivalry; and third, to provide capital to regions to 'top up' the various regional measures.

Integration, in its specific meaning here, can be defined as the combination of parts into a whole. It is obvious that it is necessary to differentiate between the immediate effects and the long-run effects when evaluating the effects of any economic integration. Immediate effects can be derived from the opportunities for rationalisation, economies of scale, specialisation and centralisation within the whole market of the enlarged European Community.

On the other hand, there are long-run effects which have a particular significance for regional issues. First, to increase specialisation and bring prosperity to some regions by stimulating greater efficiency favouring the most efficient enterprises. Nevertheless, these effects can also aggravate the performances of backward areas. Second, the efforts leading to an equalisation of factor prices, such that capital and labour flow from locations where earnings are low to those where earnings are high, should continue until an optimal position is reached. However, the level of factor earnings will be limited by the enterprises in the developed regions. Third, there might be a tendency to increase the centralisation of economic activities in the developed areas. Nevertheless, these effects will probably result in the replacement of center-periphery disparities within wider community center-periphery disparities (Glasson, 1974).

Moreover, it is argued that, '*... while both peripheral and central regions recorded falling proportions of employment in manufacturing and rising proportions in service industries during 1970s, the decline in peripheral regions' relatively small manufacturing sector was in fact faster than the decline in central regions.*' (Knox, 1984, p. 13). In fact, the periphery's growth in services such as transport, telecommunications, finance, insurance and banking was relatively slow for producer services. Conversely, the growth of education, health, public administration and retailing was relatively fast for consumer services. Meanwhile, it is observed that '*...regional variations in people's perceptions of their well-being do not demonstrate any straightforward relationship to the various 'objective' dimensions of regional prosperity ...*' (Knox, 1984, p. 14).

On the other hand, according to Ziebura (1982), although the Common Market has achieved an unprecedented rise in manufacturing goods the commercial integration of different countries is still very unequal. Accordingly; besides the increase in protectionist pressures which are the result of the national neo-mercantilist policies' contribution towards strengthening the role of coordinated economic planning. The growing lack of homogeneity of national economic policies is a response to the differentiation of the economies. According to Morgan Sant (1987), a number of theoretical statements can be derived from certain observations about the distribution and magnitude of regional disparities which vary according to the stage of development of the area, the process of 'cumulative causation', the industrial composition of the regions concerned relative location factors and social and institutional factors in the area.

Concerning interregional trade and economic integration, Bröcker (1988, p. 262) points out that '... *space has to enter into the analysis in two respects: First, different countries are not only separated by artificial trade barriers like tariffs, but also by natural barriers to trade giving rise to costs of transportation and communication. Second, integration may affect different regions of a country in a different way, thereby easing or aggravating regional disparities in a country*'. Although there are a great variety of factors influencing the costs of trading commodities between regions, Bröcker argues that simple geographical distance is clearly the most important explanatory variable of interregional trade patterns.

The role of regional disparities on integration requires discussion of the development levels of the countries concerned. Firstly, to what extent is the critical minimum effort for economic development needed for a developing area? It was suggested earlier that the 'critical minimum effort' theory is not designed to deal with the structural changes in the long-run necessary for a 'take-off' stage (Myint, 1967). Secondly, as also verified earlier, to achieve higher income levels for an economy requires the development of a certain number of regional centers, in other words, growing points in the course of the development process. In fact, as pointed out by Hirschman (1965, pp. 183-184), '... *international and interregional inequality of growth is an inevitable concomitant and condition of growth itself.*' Finally, we cite Sant (1987, p. 38), '... *regional disparities in per capita incomes and other economic and social indicators are inevitable. ... Different countries have developed different policies to meet them. ... The art of formulating regional policies lies in understanding the nature and causes of disparities and in being able to predict the impact which will follow their implementation.*'

10.2 Turkish Regional Disparities

Regional planning and development in Turkey was a new field at the beginning of 1960s when the First National Development Plan was issued and based fundamentally on sector studies and surveys of a number of branches of activities without recourse to area development methods. Since 1963, all the five-year planning periods have adopted this approach except G.A.P. [South-Eastern Anatolia Project] approach in recent years. In fact, there might be certain limitations to realising the utility of areal development. Firstly, in the specific conditions of a developing country, there are conflicts between the goals of maximising GDP on the one hand, and reducing regional disparities on the other. Secondly, there are certain limitations to using planning techniques involving regional and interregional interactions. Thirdly, there are considerable political pressures against increasing overall productivity by the rational allocation of resources to regions with high economic and social potential. In fact, as argued by Hirschman (1965, p. 109), the '*lack of interdependence and linkage is the most typical characteristic of an underdeveloped economy.*'

According to Hirschman (1965), the most influential characteristic of an underdeveloped economy is the dispersal of funds widely over the territory in the allocation of public investments. In fact, economic policies affect the rates of growth of different parts of a country by the regional allocation of public investments. Three patterns can be observed as important: dispersal, concentration on specific areas, and promoting the development of backward areas. In fact, for a long period of time the dispersed pattern has been dominant in Turkey, while the second and the third patterns are alternative political choices.

For the initial period, the problems were considered from a number of points of view: first, regions of potential development; second, backward regions; and third, metropolitan areas. The second type has a high priority in order to promote equality of benefit from the distribution of public services. On the other hand, there were two other approaches in defining regions, programming regions and functional regions. The first one comprises 8 and the second 16 regions.

The concept of the region has been used in Turkey since the early 1960s as a 'formal region' or 'programming region'. These were defined in terms of certain criteria with respect to homogeneity and reflected a simple agrarian economy for the purpose of development planning for areas displaying some economic coherence. By the early 1980s, presumably reflecting a transformation stage to a complex industrial system, 'functional region' or 'nodal region' has been used for heterogeneous units such as rural and urban settlements. These units are concerned with the interaction of the parts of certain functions and have been derived from a comprehensive analysis in terms of central place theory. Accordingly, any analysis of regional disparities using these two concepts should show some of the effects of regional differentiation on integration. In other words, they should illustrate the regional income variations concerning the transformation from an agrarian economy to an industrial system in the country.

In this contribution, the problem of getting Turkey ready for the European Community has been determined as the effects of regional differentiation on integration to EU. Accordingly, the relative structural advancement of the country for membership to the European Community may be tested by comparison to other peripheral South-European countries, namely, Greece, Portugal and Spain. Concerning data from the 1980s, one can say that the four South-European countries have achieved a relatively significant structural progress among the other developing countries, while, amongst these, Turkey is the most relatively advanced (Yilmaz, 1988; Önder and Ozmucur, 1989).

Nevertheless, not all the projects for areal development involved during the 1960s and 1970s have been implemented effectively. For example, the inability of planning to recognise the interactions among the various factors involved. In fact, recent changes in the distribution of GDP in the country, and the equipotential lines which have been transformed during the last decades, provide arguments for the prospectus for regional development in the country. In other words, the distributions of GDP during the last decades by provinces show certain regional disparities and sequential equipotential lines indicate a similar gradual advancement (Atalik, 1988).

In fact, there has been a social transformation in Turkey because the demographic and socio-economic conditions in 1960 are significantly different from the same conditions in the 1980s and 1990s. Although this transformation, which can be observed in major urban areas, seems a necessary condition to arrive at a wealthy stage, it does not comprise a sufficient condition for regional disparities (Atalik, 1988).

10.3 Results: Analysing Regional Disparities

The regional development potential approach has been selected to analyse regional disparities from a number of alternative approaches, namely, the neoclassical approach, the export-base approach, the polarisation hypothesis approach, the social overhead capital approach, and the mesostructure approach (CEC, 1986). 'The basic rationale underlying the approach adopted by the Study Group is that regional disparities are in the nature of a long-term type of problem, and not the result of short-term cyclical fluctuations. As a consequence, the emphasis both of analysis and of policy actions must be on the supply or capacity side and not on the demand side of a regional economy (see CEC, 1986, p. 62).

With respect to European Community enlargement and integration, two methods are used to examine regional development potential in the context of the relevant regional disparities: first, methods using variation coefficients to test the extent of regional disparities; and second, regression models estimating GDP variations as a function of activity rates, infrastructure, shares of industrial employment and agglomeration variables. The latter uses a Cobb-Douglas type multiple regression model to measure actual potentialities among the regions concerned.

The coefficients of variation for regional income disparities were measured for the years 1975 and 1985 respectively. For the regions based on the programming divisions, the coefficient of regional income variation moved from 0.25 in 1975 to 0.34 in 1985, while for the regions depending on the functional divisions the same coefficient moves from 0.32 in 1975 to 0.43 in 1985 (see Fig. 10.1). Thus the variability of regional income has increased over this time interval.

On the other hand, the regional development potential approach is used comparatively in the European Community studies to measure the relationships concerned (see CEC, 1986). In this analysis, in parallel to the European Community studies, a multiple regression model is used to estimate GDP variations by regions. This model is given by

$$Y = a\,L^b\,I^c\,S^d\,A^e$$
$$(\log Y = \log a + b \log L + c \log I + d \log S + e \log A)$$

where Y shows GDP by region; L, activity rate [share of the active population] by region; I, infrastructure rate [share of the literate population] by region; S, industrial employment rate by region; and A, agglomeration rate [share of the largest city] by region. The analysis was made separately for the two definitions of 8 and 16 regional divisions in the country (see Table 10.2 and Table 10.3).

Table 10.1 Regional divisions in Turkey: Programming and functional regions [1985]

Regions	Provinces
(a) Programming Regions	
1. Marmara	Balikesir, Bilecik, Bursa, Çanakkale, Edirne, Istanbul, Kirklareli, Kocaeli, Sakarya, Tekirdağ
2. Aegean	Aydin, Denizli, Izmir, Manisa, Muğla, Uşak
3. Southern Anatolia	Antalya, Burdur, Isparta
4. Çukurova	Adana, Hatay, Içel
5. Central Anatolia	Afyon, Ankara, Çankiri, Çorum, Eskişehir, Kayseri, Kirşehir, Konya, Kütahya, Nevşehir, Niğde, Sivas, Yozgat
6. Western Black Sea	Bolu, Kastamonu, Zonguldak
7. Eastern Black Sea	Amasya, Artvin, Giresun, Gümüşhane, Ordu, Samsun, Sinop, Rize, Tokat, Trabzon
8. Eastern Anatolia	Adiyaman, Ağri, Bingöl, Bitlis, Diyarbakir, Elaziğ, Erzincan, Erzurum, Gaziantep, Hakkari, K. Maraş, Kars, Malatya. Mardin, Muş, Siirt, Ş. Urfa, Tunceli, Van
(b) Functional Regions	
1. Adana	Adana, Hatay, Içel
2. Ankara	Ankara, Çankiri, Çorum, Kirşehir,
3. Bursa	Bursa
4. Diyarbakir	Diyarbakir, Bitlis, Hakkari, Mardin, Siirt, Van
5. Elaziğ	Elaziğ, Bingöl, Tunceli
6. Erzurum	Erzurum, Ağri, Erzincan, Kars, Muş
7. Eskişehir	Eskişehir, Bilecik, Kütahya
8. Gaziantep	Gaziantep, Adiyaman, K. Maraş, Ş. Urfa
9. Istanbul	Istanbul, Bolu, Çanakkale, Edirne, Kastamonu, Kirklareli, Kocaeli, Sakarya, Tekirdağ, Zonguldak
10. Izmir	Izmir, Afyon, Antalya, Aydin, Balikesir, Burdur, Denizli, Isparta, Manisa, Muğla, Uşak
11. Kayseri	Kayseri, Nevşehir, Yozgat
12. Konya	Konya, Niğde
13. Malatya	Malatya
14. Samsun	Samsun, Amasya, Giresun, Ordu, Sinop, Tokat
15. Sivas	Sivas
16. Trabzon	Trabzon, Artvin, Gümüşhane, Rize

(a) Programming Regions

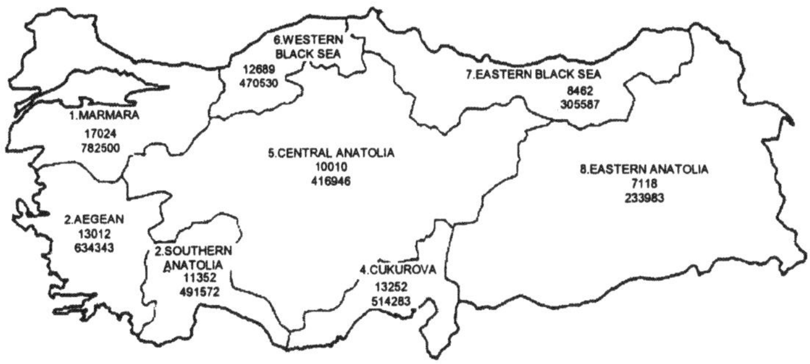

1975
$\bar{x}$= 11,616
σ= 2,912
V= 0.25

1985
$\bar{x}$= 481,218
σ= 162,648
V= 0.34

(b) Functional Regions

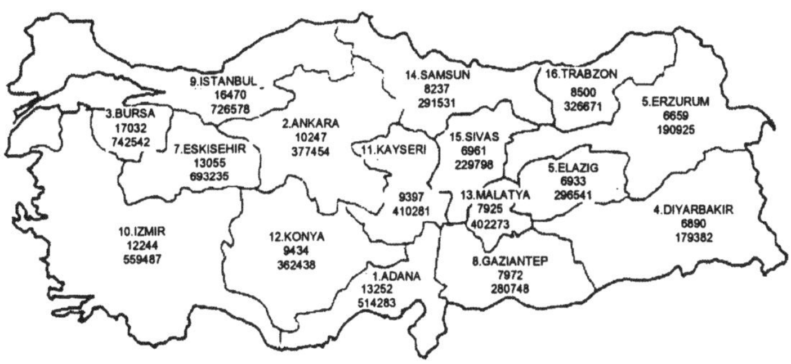

1975
$\bar{x}$= 10,094
σ= 3,253
V= 0.32

1985
$\bar{x}$= 411,508
σ= 178,865
V= 0.43

Fig. 10.1 GDP per capita [TL] and coefficient of regional income variation [V], 1975 and 1985

Table 10.2 GDP variations by regions as a function of activity; infrastructure, industrial employment and agglomeration rates in Turkey [1985] [Programming Regions]

Regions	log Y	log L	log I	log S	log A
(1) Marmara	2.8937	1.7951	1.9237	1.3096	1.6928
(2) Aegean	2.8020	1.8115	1.9047	1.3053	1.4297
(3) Southern Anatolia	2.6918	1.8228	1.9185	1.2671	1.2355
(4) Çukurova	2.7109	1.1774	1.8920	1.3117	1.3159
(5) Central Anatolia	2.6101	1.8055	1.8926	1.1815	1.3074
(6) Western Black Sea	2.6130	1.8135	1.8825	1.3074	0.7708
(7) Eastern Black Sea	2.4857	1.8432	1.8744	1.0718	0.6532
(8) Eastern Anatolia	2.3692	1.8401	1.7810	1.0755	0.6627

Table 10.3 GDP variations by regions as a function of activity, infrastructure, industrial employment and agglomeration rates in Turkey [1985] [Functional Regions]

Regions	log Y	log L	log I	log S	log A
(1) Adana	2.7109	1.1774	1.8910	1.3117	1.3159
(2) Ankara	2.5763	1.7965	1.8937	1.1613	1.7032
(3) Bursa	2.8709	1.1535	1.9211	1.4638	1.6655
(4) Diyarbakir	2.2528	1.8481	1.7007	0.9912	0.9867
(5) Elaziğ	2.4127	1.8401	1.8155	0.9493	1.3180
(6) Erzurum	2.2810	1.8603	1.8155	1.1238	0.9684
(7) Eskişehir	2.8407	1.7895	1.9216	1.3263	1.4502
(8) Gaziantep	2.4487	1.8115	1.7958	1.1613	1.1986
(9) Istanbul	2.8615	1.8014	1.9143	1.4232	1.7092
(10 Izmir	2.7474	1.8169	1.9074	1.2718	1.2380
(11) Kayseri	2.6127	1.8202	1.8870	1.1818	1.3463
(12) Konya	2.5587	1.8088	1.8790	1.1522	1.2764
(13) Malatya	2.6042	1.7951	1.8585	1.1958	1.5622
(14) Samsun	2.4653	1.8475	1.8662	1.0253	0.8129
(15) Sivas	2.3617	1.8135	1.8633	1.1003	1.4099
(16) Trabzon	2.5145	1.8363	1.8864	1.1303	0.9294

The estimated multiple regression equations are

$$\log Y = -2.68 + 1.30 \log L + 0.87 \log I + 0.91 \log S + 0.18 \log A,$$

where

$$\begin{aligned} \log\ a &= -2.68 \\ b &= 1.30\ (+0.73) \\ c &= 0.87\ (+1.13) \\ d &= 0.91\ (+2.38) \\ e &= 0.18\ (+1.87),\ \text{S.E.} = 0.054,\ R^2 = 0.95, \end{aligned}$$

for the division of programming regions; and,

$$\log Y = 0.81 - 0.78 \log L + 1.27 \log I + 0.62 \log S + 0.05 \log A,$$

where

$$\begin{aligned} \log\ a &= 0.81 \\ b &= -0.78\ (-0.46) \\ c &= 1.27\ (+2.27) \\ d &= 0.62\ (+2.11) \\ e &= 0.05\ (+0.38),\ \text{S.E.} = 0.088,\ R^2 = 0.85, \end{aligned}$$

for the division of functional regions. The figures in brackets represent the t-statistics for the partial regression coefficients.

The figures show that activity rate [*L*], share of industrial employment [*S*] and infrastructure rate [*I*] all give statistically significant explanations for regional GDP variations. The agglomeration rate indicator [*A*] gives a slightly less than significant explanation, which may be due to the fact that GDP might depend more on population size than on the level of development at a stage of pseudourbanisation. One may derive from these results the conclusion that improvements in activity, industrial employment and social welfare infrastructure are conditions of regional development, because both of the two divisions have high multiple correlations. Moreover, the correlation matrices and parameter estimates show considerable performance in comparison to the European Union figures (see CEC, 1986).

10.4 Conclusions

The findings about the European Community and regional disparities should not be decisive for the decision making processes in the EU regarding enlargement and integration. It may be said that if functional regions are indispensable for regional growth in an area, then regional disparities will be inevitable in the European Community.

In the Turkish case, the regions depending on the programming divisions show higher relationships than the regions depending on the functional divisions, perhaps reflecting sample size, while the variation coefficients for regional income give higher figures for the functional regions. In fact, multivariate analysis indicates the strong relationship of GDP variations by regions as a function of activity, infrastructure, industrial employment and agglomeration rates in addition to higher variation coefficients in regional income. Consequently, potentialities should be fundamental to achieving higher performance towards integration in the European Community.

The quasiproduction function based on a Cobb-Douglas specification gives significant explanations for the specific regional determinants on activity, social infrastructure, industrial employment, and agglomeration rates similar to the EU factors. The results show that the various potentiality factors are significant for testing the relevance of the regional development potential theory.

References

Atalik G. (1988): Spatial Interaction Process in a Social Transformation Stage: The Turkish Urbanisation. Paper Presented at the 28th European Congress of the Regional Science Association, August 1988, Stockholm

Bröcker J. (1988): Interregional Trade and Economic Integration: A Partial Equilibrium Analysis, *Regional Science and Urban Economics* 18, 251-281

Commission of the European Communities (1984): *The Regions of Europe*, Office for Official Publications of the European Communities, Luxembourg

Commission of the European Communities (1985): *The Contribution of Infrastructure to Regional Development*, Office for Official Publications of the European Communities, Luxembourg

Eurostatistics (1988): *Europe in Figures*, Office for Official Publications of the European Communities, Luxembourg

Glasson J. (1974): *An Introduction to Regional Planning*, Hutchinson Educational, London

Hirschman A.O. (1965): *The Strategy of Economic Development*, Yale University Press, New Haven and London

Knox P.L. (1984): *Regional Disparity and Regional Change in the European Community*, [Coursework Supplement No.3], Department of Geography, University of Dundee

Myint H. (1957): *The Economics of the Developing Countries*, Hutchinson University Library, London

Myrdal G. (1972): *Economic Theory and Underdeveloped Regions*, Methuen, London

Önder I. and Özmucur S. (1989): Turkish Economy in the Year 2000. In: Turkish Political Science Association (ed) *Turkey in the Year 2000,* Sevinç Matbaası, Ankara, pp. 133-154

Sant M. (1987): *Regional Disparities*, Macmillan, London

Yilmaz T. (1988): *The Relative Structural Advancement of Turkey in the Context of Being Full-Members of the EEC*, Iktisadi Kalkinma Vakfi Yayinlari, Istanbul

Ziebura G. (1982): Internationalization of Capital, International Division of Labour and the Role of the European Community, *Journal of Common Market Studies* 12, 127-146

11 An Institutional and Cultural Perspective on Romanian Regional Development Policy

Daniela L. Constantin
Academy of Economic Studies of Bucharest, Romania

11.1 Introduction

The regional dimension of the transformation processes undertaken in Central and Eastern European countries represents a new field of research in regional science (van Geenhuizen and Nijkamp, 1995a). The problems involved in the restructuring of regional economies, the question of regional policy instruments in a context of acute shortage of financial means, the question of decentralisation, the impact of European integration etc. represent central issues in analysing the regional economies in these countries in transition.

This chapter focuses on the institutional and cultural aspects of regional development policy, with special emphasis on Romania. The starting point is the fact that the regional dimension of the transition has to be integrated within a framework which combines the need for local identity, self-reliance and development with the challenges and opportunities of globalisation processes on one hand and the perspective of future integration into the European Union on the other. Accordingly, any project in the field of regional development is multifaceted by nature. On one hand it focuses on the issues of regional economics and policies in order to assess the actual situation of regional development, to identify appropriate objectives and to derive there from realistic economic and social policies, with corresponding instruments and mechanisms of implementation and evaluation. On the other hand it has to concentrate on the institutional framework and allocation of responsibilities within the organisational structures of public administration.

The chapter is divided into two major parts. Section 11.2 views the institutional reforms in Romania at a regional level after 1990, as a component of the process of transition on the path towards a democratic society. Section 11.3 addresses the cultural and behavioural challenges involved in order to implement regional policy, and attempts to illustrate that participation, communication, mentality, attitudes – even more than financial constraints – are crucial elements for the real success of the implementation of regional policies.

11.2 Regional Development Policy in Romania: Institutional Reforms

Despite the undeniable importance of the success of the process of transformation, the regional dimension has been neglected for many years in Romanian policy. Only in 1995, when faced with having to prepare a strategy for accession to the European Union did the Romanian government recognize that the problems of regions and local communities represent key elements, to be necessarily taken into account in order to bridge the gap between words and facts in the debates on decentralisation and local administrative autonomy on one hand, and those on European integration, transborder co-operation, spatial networks, etc. on the other. Accordingly, two special chapters of the afore-mentioned strategy have focused on regional issues, representing the background of subsequent decisions and actions. These chapters refer to regional development strategies and to the national spatial plan, concentrating on large-scale infrastructure projects and rational land use, as a synthesis of the strategies developed by each county and by the Bucharest municipality. They combine the aspects concerned with transition and reform processes at a regional level with those actions that have to be undertaken in order to achieve integration with structures of the European Union.

The development of this strategy has been followed by a programme for regional policy analysis and development established within the framework of PHARE by the European Union and the Romanian Government. The programme has been implemented by the Department of Local Public Administration of the Romanian Government, assisted by a team of foreign advisors. A Regional Policy Task Force supervised the activities of the programme with representatives from key ministries and regional authorities. One of the main tasks of this programme has been the preparation of a 'Green Paper', including the proposals formulated by the Task Force according to which the government should design and implement regional policy in Romania. The proposed policy has three essential objectives: *First,* to prepare Romania for EU membership and to get access to the EU structural funds; *second,* to reduce regional disparities among Romanian regions; *third,* to integrate public sector activities in order to enable the regions to reach a higher level of development (Green Paper, 1997).

Of course, the challenges of transition at the national level – a new institutional framework, privatisation, restructuring and technological change, new entrepreneurial relationships, etc. – specifically mirrored at the regional level, must be central issues in the Romanian regional development strategy and corresponding regional policies. On the other hand, as transition is not an isolated purpose in itself, the strategy conceived for this period should take into account not only the objectives specific to this stage, but also the long-term goals, expressing the time-continuity of strategic choices. This means that the national and regional contexts have to be integrated within a complex framework, able to take into account crucial issues such as globalisation and regionalisation, European integration, worldwide ecological crises and socio-political changes (Thierstein and Egger, 1995).

For Romania, a country which has made remarkable contributions to European culture, the possibility of joining the European Union is a major challenge. This explains why many Romanians see only one side of the coin, namely the advantages of integration that do not need to be discussed here. It is more important for the policy-makers to have a clear understanding of the disadvantages of an insufficient and/or inadequate preparation of this process, as well as to recognize the key elements able to shape the relationships between Western and Eastern Europe in such a way that makes integration a success story for both the West and the East (van Geenhuizen and Nijkamp, 1995b.). Even according to many Western European researchers '*a reliable path to reintegration should be based on the basic needs of East and Central Europe and should not be guided by that which is still lacking within the EU. Key elements in this process could be: assistance in the development of a democratic institutional framework based on the political heritage of these countries; market access for products and industries in which Eastern European countries have comparative advantages [such as agriculture, textile and certain sectors of manufacturing]; the transfer of knowledge and technology to the industrial base of the receiving regions in order to avoid the creation of isolated development poles unlinked to the surrounding society*' (Cornett, 1994, p. 12).

As far as Romania's particular situation is concerned, attention should be concentrated on its relative advantages in a very dynamic environment [its size and strategic geopolitical location, the variety and quality of raw materials and, especially, its human resources, the economic potential of agriculture, certain industries, tourism, etc., the political and social stability, and so on]. On the other hand the policymakers should be aware of the drawbacks which are the accumulated result of nearly fifty years of centrally planned production and amplified by the difficulties of the transition process. In a period of unprecedented openness between countries and regions, which increases the competition between European regions in an emerging network economy, special emphasis should be put on the removal, within a reasonable time limit, of the main barriers to transborder co-operation [physical, economic, political and socio-cultural].

The most significant conclusion that can be drawn so far is that, despite the importance of international support, internal efforts and commitment to building a new society remain the basic requirements for creating a competitive economy, regionally oriented, that can allow for the obvious tendency towards decentralisation specific to the transition phase on one hand, and the clear option for integration into a Europe of regions on the other. An integral market-oriented regional strategy and policy [Thierstein and Egger, 1995; Hiadlovska and Vogd, 1994) could be conceived as a corner stone in accomplishing this goal. In general terms, the main objectives would be to reduce regional unemployment, to attain an efficient geographical distribution of industry and employment and, last but not least, to provide a more equal geographical distribution of income and living conditions (Hansen et al., 1995).

From an overall point of view, a regional policy able to carry out these objectives should combine the efforts, at all levels, involved in promoting regional development, should concentrate on the actors and their behaviour, should coordinate sectoral policies and environmental preservation in accordance with the complex relationship between these and spatial organisation, and should strengthen cooperative problem-solving instruments. This policy can directly influence regional competitiveness, and contribute to the augmentation of the potential of a region, in order to ensure high economic development levels and living standards.

The question of the levels involved in regional development is closely related to the renewal of the institutional and legal framework, seen as a crucial element of the Romanian structural reform. Within the clear tendency to decentralisation, the regional strategies and policies focus on regional [local] efforts to foster socio-economic development, taking into account the strengths and weaknesses of each county.

But '*viewed in the context of an unexpected intensification of spatial polarisation tendencies in East European countries, local authorities, local initiatives and local efforts and their role in regional development policy must not be considered as an alternative to the policy conducted from the perspective of the spatial organisation of the whole country or of a group of countries. Both national and even supranational and local regional policies must be complementary to one another*' (Böttcher et al., 1991/1992, p. 25). This complementarity is closely related to the scale, importance and particularities of each project which has a spatial impact. For example, in a complex network economy, large-scale infrastructure projects, which always more frequently cross regional and international borders, need co-ordination between regional policies at a national level and those at a county [local] level.

Despite the declared formulation of local self-administration, Romanian local authorities and communities were treated as voiceless executors of commands from the central level until 1996. Therefore, empowering the local, democratically elected authorities, providing them with legal, financial and other instruments of basic action within their counties and towns, is a necessary component of the basic social and economic reform, highlighting the tendency to decentralisation. At the same time the other levels [national, international] of regional policy must also be considered, as a response to the subsidiary principle, according to which decisions must be taken at the most appropriate spatial level (see van Geenhuizen and Nijkamp, 1995a).

These overall requirements have been taken into account by the main institutional, legislative, and administrative acts that have a regional impact in Romania. The Local Elections Act, the Local Public Administration Act, the Regional Development Act, the Local Taxes Act, the Local Budget Act and the Act of State and Local Community Patrimony are the most important examples in this respect. After 1989 Romania signed the European Charter of Local Autonomy, issued by the Council of Europe and developed a regional dimension as preparation for the accession of Romania to the European Union.

The law regulating local elections [the Local Elections Act] was one of the first regulatory measures and enabled the first local elections to take place under the new democratic regime in 1992. Thus, in each constituency coinciding with the county and town boundaries, the citizens vote simultaneously, but separately for the county councils and town councils, for a four-year mandate. The elections are held by direct universal suffrage under a list system of proportional representation, where seats are allocated in accordance with the Hondt highest average method. This has been a remarkable achievement for a country where a few years previously the parliamentary and local elections were simple 'simulations', almost all candidates receiving at least 90 percent of votes.

Another major step in the direction of implementing the local administrative autonomy was the construction of a basic legal framework for the local authorities: the Local Public Administration Act, that defined the main points regarding the local government system. These points refer to the allocation of functions, the organisation of local units, the relationship between levels of government, and the financial framework. In fact, this act has represented the practical basis for the development of local autonomy. Concretely, the power of the basic local authorities mainly refers to local development programming, spatial planning, the supply of public services [such as public transportation, roads, water supply, sewage disposal, refuse collection, etc.], financial participation in public/private projects in order to carry out actions or work of common interest, investments of local interest, trade fair organising, building and demolition permits, nature conservation and site protection, etc. Parliament also issued the Local Taxes Act, that regulates the sources of local funds and their destinations in detail.

Even though these acts seemed to be encouraging for the success of the implementation of the principles of local autonomy, the facts still showed a quite important gap between words and action in this field. Many mismatches and drawbacks continued to exist. For example, various reasons [economic, political, etc.] have made it difficult for local taxes to be effective. Their revenue base is still rather weak. Moreover, there was not a clear, systematic view of the real instruments which might be used to implement certain regional policies in economic practice. The specific tasks established for the local public administration institutions had to be revised too, so as to avoid overlaps, interferences and contradictions in terms of the goals to be achieved in certain fields of activity and the tasks of the institutions involved therein.

The November 1996 elections brought the former opposition into power, and a set of priority actions aiming to correct and complete the legal institutional framework of local public administration were established. The government has focused on the need for a real decisional, financial and patrimonial autonomy of county and town administrations, establishing new ways of improving the system, ameliorating the relationships between different levels of government, enlarging the scope of local action, and finding the legal forms which can give coherence and public unity to local actions in the field. Accordingly, the Local Public Administration Act has been modified and supported by two new acts: the Local Budget Act and the Law of State and Local Community Patrimony. All these measures follow the scheme of a relative autonomy model that gives independence to local authorities, without denying the reality of the nation state. The emphasis is on giving freedom of action to local authorities within a defined framework of power and duties. Central government relationships with local authorities are therefore determined largely by legislation. Control is limited. Within the relative autonomy model local authorities may pursue policies which they share with central government or which differ from those advocated by central government in some cases. The final result should be the so-called 'reinforcement of local power', making the local administration able to act freely within a legally pre-determined framework. In this respect, the existing national association of municipalities could become a consistently demanding body, an active lobby, formally consulted in most government decisions regarding local authorities.

But local autonomy must not be seen in absolute terms. The co-operation between central and local administration for carrying out programmes of national interest or local projects exceeding the county/town-funding capacity should also be considered, as well as the need for co-ordination between local authorities with regard to their development policies. This coordination is necessary since the benefits of such policies will spill over into neighbouring counties/towns and '*acting independently will lead to under-funding of local development, unless the free-rider problem can be overcome*' (Armstrong and Taylor, 1993, p. 24). The co-ordination between local authorities can create the basis for local economic development partnerships, possibly to be combined with public-private partnerships in local co-operative networks. West European and U.S. experience offers interesting lessons, like that on intergovernmental units [defined as combinations of counties that rarely finance, but frequently administer local development programmes] and special districts [units that include a specific population utilising a particular service; examples are an airport district, a port authority or a school district] (Leven, 1990).

These overall requirements, important for the improvement of the administrative framework for a decentralised regional policy, were reflected in the Green Paper (1997) that proposed a specific institutional structure for regional development policy. Then they were included in the Regional Development Act, issued in 1998 (Law 151, 1998). According to this act eight development regions have been created to serve as the framework for conceiving, implementing and evaluating regional development policy as well as for collecting the statistical data corresponding to the NUTS2 level of the EUROSTAT. Each region includes up to seven counties, associated on a voluntary basis [the total number of counties is 41, plus Bucharest municipality, corresponding to the NUTS3 level].

In institutional terms, the central unit, responsible for national regional development strategy and the implementation of regional development policy is the National Council for Regional Development [NCRD], subordinated to the Government of Romania. In short, the tasks of this institution are: the elaboration of the National Regional Development Programme; to make proposals to the government regarding the amount of the National Regional Development Fund [NRDF] and its financial sources; to take decisions regarding the allocation of funds from the NRDF to various Regional Development Funds [RDFs]; to administer the NRDF and to monitor the RDFs; to make proposals for legislation and new regulations in regional development; and to stimulate co-operative actions at a regional level in order to reduce regional disparities and to diminish or eliminate the specific problems of some critical areas. The executive, operational body of the NCRD was, in the beginning, the National Regional Development Agency [NRDA]. In January 2001 the NRDA was transformed into the General Regional Development Directorate [GRDD] of the Ministry of Development and Forecasting, newly created by the present government.

The idea of decentralising some of the responsibilities for regional development policy to the regional/local level led to the establishment of regional councils and regional agencies, with specific responsibilities towards the NCRD and NRDA/GRDD on one hand, and towards the region on the other. The latter responsibilities refer to the defining and implementing of the regional development plan; obtaining from the NCRD the funding necessary for the implementation of regional development programmes; managing the RDF with regard to tasks that can be best solved at a local level.

However, this new institutional framework has an important weakness due to the fact that the eight regions 'are not administrative-territorial units and do not have judicial bodies' (Planul National de Dezvoltare, 1999, p. 7). Moreover, the regional development agencies act as non-governmental organisations (Law 151/1998) and this creates difficulties in a normal relationship between the NRDA/GRDD and these agencies. Therefore additional transformations are required, to make the development regions work effectively and really serve as a basis for decentralising regional policy and implementing the programmes with which structural funding is concerned.

Following the creation of the regional development institutional structure, the NRDA elaborated the National Plan for Regional Development, as a major component of the National Development Plan [NDP] in 1999 [the first version, updated in the following years]. The NDP has been conceived as the planning instrument for achieving the complex objective of economic and social cohesion and for ensuring the coherence and compatibility with the planning and programming documents employed by the EU member countries. The National Plan contains the strategic priorities for the period 2000 – 2006, as well as the development priorities and implementation measures for 2000 – 2002, referring to: private sector development and the supporting investment process; supporting productive SME development; regional and local infrastructure improvement and development; human resources development; tourism development; supporting technological development and innovation; agriculture and rural space development; transport infrastructure development; environmental protection. The financial assistance from the EU for supporting these priorities mainly comes via PHARE, ISPA and SAPARD programmes.

As far as regional development planning is concerned – at both national and regional levels – it is still too sectorally oriented, leading, in many situations, to an ineffective distribution of development funds. A solution would consist in changing the emphasis from the sectorally oriented approach to a complex area programme-project package approach, aiming to turn to good account the endogenous development potential (Constantin, 2001). Each programme, with its corresponding projects, should focus on a top priority development domain in the envisaged area, able to create major employment and income multiplier effects, so as to stimulate, via indirect and induced effects, other economic and social sectors as well. In most cases, the effects will spill over into the neighbouring area – belonging to the same region or to different regions, offering the support to intra and interregional co-operation, within a 'generative' growth model. As the experience of many Western countries has demonstrated, this model brings about many advantages compared to the 'competitive' growth model, applied in a centralised context. This view is supported by many Romanian regional scientists, working in both the academic, research field and in various administrative structures (e.g. Constantin 2001, Nica, 2001 and Suciu, 2001).

11.3 Cultural and Behavioural Challenges in the Implementation of Regional Policies

Undoubtedly, defining objectives, planning the strategy, and proposing economic policy actions are major components in the strategy process, but the most difficult time for most strategies is when plans must be turned into action. International experience shows that many strategies have not successfully made the transition, while others have been only partially implemented. For some, the strategy process appears to have stopped dead following the preparation of the main document (Carrew-Reid et al., 1994).

The behavioural challenges – participation, communication, mentality, and attitudes seem to be crucial elements for the real success of regional strategy and policy, even more than financial constraints. A 'people-centred' approach to regional development strategy can result in a more realistic strategy, with a broader base of knowledge, understanding and commitment from the groups involved and with better links to promising local initiatives (Blunt, 1995). All actors involved can be identified in this phase: national government, local governments, the corporate sector, non-governmental organisations and, last but not least, communities and individuals. Thinking that participation is entirely a non-governmental affair would be a big mistake. Government is that actor that can help to provide the right conditions for participation (Carrew-Reid et al., 1994).

In a democratic political system local government is a key element. In general terms, local government organisations provide a range of services which are central to the social and economic well-being of their citizens. They are large-scale spenders of public money and major sources, both directly and indirectly, of employment. But the local administration is more than just a supplier of services. It is the level of government closest to the citizens and has a role in representing the concerns and views of the locality (Batley and Stoker, 1991). It should build structures and create an environment which can execute policies in order to actively support participation.

The actions concentrated on local administration reform, as identified in the previous section, cannot be considered complete without a real commitment to the renewal of the corresponding organisational structures, in order to essentially transform the existing mechanistical structures into organic ones, more flexible, more effective and certainly less bureaucratic. A KPMG survey reveals that the major barriers perceived by foreign investors in Romania are stifling bureaucracy [71 percent], poor infrastructure [60 percent], and corruption [55 percent]. Accordingly, there is an obvious need to change the civil servant's status and behaviour on one hand, and the mentality and behaviour of the local communities on the other hand, so as to make their involvement in promoting local development projects possible. Indeed, the modern approach in analysing local government is a behavioural approach. Behaviour simply means what people do, as opposed to what they say they do or what they are supposed to do, in legal and institutional terms (Stayner, 1980).

Thus, democracy depends on democratic values as well as on organisational efficiency at all levels. Democracy is a principle not just of political organisation but also of human relations, meaning that democratic reform is partly about the organisation of the civil service, but also about attitudes (Ridley, 1995). A civil service law has been issued quite recently in Romania, but it will not of itself change the way in which civil servants behave. This is the real challenge of democratisation. Of course, the reform of the legal and institutional framework is a necessary starting point, but it is only a means to an end and 'it is taking a great risk to think that is sufficient' (Ridley, 1995, p. 11).

An inquiry into the place of civil service in democratic states and the implications for civil service organisation reveals several models, but given the differences in culture, political climate and the economic situation in Central and East European countries, they can be accepted only if they are really suitable. It is not our intention to discuss these models here. However, there are some ideas of general relevance which are important and should be noted: the need to increase managerial efficiency and to make public services and their personnel more responsive to the citizens they serve; that the human element should be focused on rather than formal civil status; and motivation for good behaviour. A democracy requires democrats, so that the longer term essential in a democratisation process is not a constitutional reform but the spread of democratic attitudes, respect for the people as much as respect for law, in public service.

The other side of the coin regards the participation of local communities in local development. The concept of regional culture [and identity] is a central theme in examining this issue, based on the analysis of the forms of the sense of belonging. The identity and solidarity among people or specific groups living in a region represent a key element in promoting a 'development by the people' concept [that creates, beside development of the people and development for the people, the basis of a more comprehensive concept, that of sustainable human development] (Blunt, 1995). This view implies that people, by rights, should have access to a variety of ways of exercising power. Its logic suggests that participation is both a means and an end, and that decentralisation of public administration is desirable for its own sake. Research evidence shows that perceptions of what constitutes participation, as well as views concerning its desirability vary between cultures. This has much to do with the well-established differences that exist between cultures regarding attitudes to authority, uncertainty and group loyalties versus individual loyalties.

In Romania the long communist period had a very bad influence on the behaviour of local communities, as well as on the whole of society: the way of perceiving democratic values; attitudes have been perverted to a great extent, so that the transition to a democratic society has to face enormous challenges in order to alter wrong mentalities and attitudes (Constantin, 1997). Thus, in the presence of an alienating gap between words and facts [e.g. common, everybody's ownership equal to nobody's ownership], supported by an economic policy based on forced industrialisation, which imposed the leading role of the working class and marginalized the intelligentsia ['we work, not think'] communism induced a lack of respect for the longstanding, fundamental values and traditions, and blurred intercultural variety and regional identities.

The rural areas, affected by the policy of depopulation, and by ageing and predominantly feminine labour resources, were converted to an 'urban' lifestyle only by imitation [food, clothing, music, etc.], whereas the standard of living has remained very low, with a poor quality of basic infrastructural facilities. All of this induced serious disturbances in rural community life and decreased the capacity to benefit from the traditional advantages that many villages could normally have had.

The urban environment was altered, too: immigration, work in factories, and life in small apartments, sometimes deprived of basic comforts, the new, monotonous architecture replacing many old, invaluable monuments which were demolished over night, made people lose their sense of belonging to a community with a well-defined identity in terms of values and traditions. However, as an expression of the resistance to the communist dogmatism, a quite important number of rural areas, especially in the mountains, have been very successful in preserving their old traditions, lifestyle and fundamental values.

After 1989 many civil society organisations have been created, and the co-operation between them and the administrative structures could contribute to fulfilling this purpose. It has been demonstrated that the achievements of governments depend on the co-operation and involvement of other economic and social actors, such as 'community organisations'. Two broad types can be identified: people's organisations and non-governmental organisations. People's organisations [like mutual-assistance farming groups, urban housing, welfare associations, and women's groups, etc.] represent their members' interests, are accountable to their members, and tend to participate in organisational structures. At a local level they can make up the nucleus of the increasing participation of local communities in the promotion of development programmes corresponding to their own interests.

As regards the activity of non-governmental organisations, it would be substantially improved if a real, deep engagement with community groups and individuals were brought about: the local level is the most practical one for public participation. A relevant example is the field of environmental conservation, where the capacity of local pressure groups to determine important decisions for cleaning up the environment was very strong when large information campaigns were undertaken [e.g. the campaign for sending back to Germany a large quantity of toxic waste, imported as pesticide in an area of outstanding natural beauty, Sibiu]. However, after nearly fifty years of environmental information monopoly and lack of real education in this field, it is hard to believe that a completely new way of perceiving the environmental issues can be achieved in a few years, burdened with the difficulties of transition. Even though it cannot be denied that efforts to create a real environmental culture are under way, the economic and social problems, such as production decline and increasing unemployment, have made public concerns shift away from environmental issues to more immediately stressful ones: the falling value of real wages, increasing job insecurity.

In conclusion, Romanian society still suffers from the consequences of communism. They did not disappear at the same time with the totalitarian state. And, as a famous journalist wrote: '*We will be able to talk about the failure of totalitarianism only when our souls, our minds will be completely decollectivised. This 'privatisation' is infinitely more difficult than that occurring in the economy and, in fact, it has started only at instinctive and good will level. Under these circumstances how could one discuss post-transition?*' (Paler, 1997, p. 1)

11.4 Conclusions

The regional dimension of transition implies a specific approach, allowing for practical and theoretical aspects, in concrete time and space circumstances.

Considering the real Romanian situation, the application of regional policy in a decentralised context represents a basic condition for the success of reform at the regional level. Therefore, special emphasis must be put on the institutional framework and the allocation of responsibilities within the public administration organisational structures. At the same time, there is an obvious need to change the status and behaviour of civil servant, on one hand, and the mentality and behaviour of the local communities on the other hand, in order to make possible their involvement in the promotion of local development projects in accordance with their major interests.

References

Armstrong H. and Taylor J. (1993): *Regional Economics and Policy* [2nd edition], Harvester Wheatsheaf, New York, London

Batley R. and Stocker G. (eds.) (1991): *Local Government in New Europe. Trends and Developments*, MacMillan, London

Blunt P. (1995): Cultural Relativism, Good Governance and Sustainable Human Development, *Public Administration and Development* 15, 1-9

Böttcher H., Funck R. and Kowalski J. (1991/1992): Reflections on the Spatio-economic Consequences of the Post-Communist Transition in Eastern Europe, *Jahrbuch für Regionalwissenschaft* 12/13, 9-28

Carrew-Reid J., Prescott-Allen R., Baas S. and Dalal-Clayton B. (1994): *Strategies for National Sustainable Development*, Earthcan Publications, London

Constantin D. (1997): Institutions and Regional Development Strategies and Policies in the Transition Period: The Case of Romania, *37th Congress of the European Regional Science Association*, Rome, August 1997

Constantin, D. (2001): Some Considerations Regarding the Improvement of the Institutional-Administrative Framework of Regional Development in Romania from the Perspective of French Experience, *The First National Symposium of the Romanian Regional Science Association*, Bucharest

Cornett A. (1994): Regional Economic Integration – A Pragmatic Approach toward Reintegration of East and Central Europe, *34th Congress of the European Regional Science Association*, Groningen, August 1994

Geenhuizen M. van and Nijkamp P. (1995a): Corporate Transborder Networking in Eastern Europe: Potentials of an East-West Development Zone, *35th Congress of the European Regional Science Association*, Odense, August 1995

Geenhuizen M. van and Nijkamp P. (1995b): The Dynamics of Regional Science, *Tinbergen Institute Discussion Paper* 43/1995, Amsterdam

Green Paper (1997): *Regional Development Policy in Romania*, Romanian Government and European Commission, PHARE Programme

Hansen J., Young E. and Cornett A. (1995): Regional Development Policy in the U.S. and Europe: Lessons from Local Initiatives, *35th Congress of the European Regional Science Association*, Odense, August 1995

Hiadlovska M. and Voogd H. (1994): Marketing Planning for Regional Economic Development: A Conceptual Framework, *34th Congress of the European Regional Science Association*, Groningen, August 1995

Legea 151/1998: Privind Dezvoltarea Regionala in Romania, *Economie si administratie locala* 12/1998, 12-14

Leven C. (1990): Local Initiatives for Economic Development. Lessons from U.S. Experience, *Foundation for Local Democracy Seminar*, Warsaw

Nica A.N. (2001): Regional Development Policy in Romania: Between Internal Necessity and External Requirement Or the Philosophy of the European Support for Regional Development, *The First National Symposium of the Romanian Regional Science Association*, Bucharest

Paler O. (1997): Ce-o fi Post – Tranzitia?, *Romania Libera* 2267/9 septembrie, 1997

Planul National de Dezvoltare (1999): Agentia Nationala pentru Dezvoltare Regionala, Bucuresti

Ridley F.F. (1980): Civil Service and Democracy: Questions in Reforming the Civil Service in Eastern and Central Europe, *Public Administration and Development* 15, 11-20

Stayner J. (1980): *Understanding Local Government*, Martin Robertson, Oxford

Suciu M.C. (2001): Regional Labour Market Disparities, *The First National Symposium of the Romanian Regional Science Association*, Bucharest

Thierstein A. and Egger U.K. (1995): An Integral Regional Policy Perspective. Lessons from Switzerland, *35th Congress of the European Regional Science Association*, Odense, August 1995

List of Figures

List of Tables

Author Index

List of Contributors

Gündüz Atalik
Department of Urban and Regional Planning
Istanbul Technical University
80191 Taksim, Istanbul
Turkey

Peter Batey
Faculty of Social and Environmental Studies
University of Liverpool
23 Abercromby Square
Liverpool L69 7ZG
United Kingdom

Riccardo Cappellin
Department of Economics and Institutions
University of Rome Tor Vergata
Via di Tor Vergata
I-00133 Roma
Italy

Manas Chatterji
School of Management
Binghamton University
State University of New York
P.O. Box 6015
Binghamton, N.Y. 13902-6015
U.S.A.

Daniela L. Constantin
Department of Statistics and Economic Forecasting
Academy of Economic Studies of Bucharest
Bd. Lascar Catargiu 16A, ap.8
Sector 1
RO-71113 Bucharest
Romania

Manfred M. Fischer
Department of Economic Geography & Geoinformatics
Vienna University of Economics and Business Administration
Rossauer Lande 23/1
A-1090 Vienna
Austria

Peter Friedrich
Department of Economics and Organisational Sciences
University of the Federal Armed Forces
Werner Heisenberg Weg 39
D-85577 Neubiberg
Germany

Peter Nijkamp
Department of Spatial Economics
Free University
De Boelelaan 1105
NL- 1081 HV Amsterdam
The Netherlands

Bengt Lorendahl
Södertörns Högskola/
Södertörn University College
Alfred Nobels Allé 11
Box 4101
SE-14104 Huddinge
Sweden

Roger Vickerman
CERTE
Department of Economics
Keynes College
University of Kent
Canterbury, Kent CT2 7NP
United Kingdom

Sytsze A. Rienstra
Department of Spatial Economics
Free University
De Boelelaan 1105
NL- 1081 HV Amsterdam
The Netherlands

Jaap M. Vleugel
Department of Spatial Economics
Free University
De Boelelaan 1105
NL- 1081 HV Amsterdam
The Netherlands

Richard T.M. Smokers
Department of Spatial Economics
Free University
De Boelelaan 1105
NL- 1081 HV Amsterdam
The Netherlands

Xiao Feng
Department of Economics and
Organisational Sciences
University of the Federal Armed Forces
Werner Heisenberg Weg 39
D-85577 Neubiberg
Germany

MIX
Papier aus verantwortungsvollen Quellen
Paper from responsible sources
FSC® C105338

If you have any concerns about our products,
you can contact us on
ProductSafety@springernature.com

In case Publisher is established outside the EU,
the EU authorized representative is:
Springer Nature Customer Service Center GmbH
Europaplatz 3, 69115 Heidelberg, Germany

Printed by Libri Plureos GmbH
in Hamburg, Germany